DEBATING MATTERS

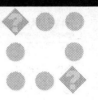

SCIENCE:

CAN WE TRUST THE EXPERTS?

Institute of Ideas
Expanding the Boundaries of Public Debate

Tony Gilland
Sue Mayer
Bill Durodié
Ian Gibson
Douglas Parr

Hodder & Stoughton

A MEMBER OF THE HODDER HEADLINE GROUP

DEBATING MATTERS

Orders: please contact Bookpoint Ltd, 130 Milton Park, Abingdon, Oxon
OX14 4SB. Telephone: (44) 01235 827720. Fax: (44) 01235 400454.
Lines are open from 9.00–6.00, Monday to Saturday, with a 24 hour message
answering service. Email address: orders@bookpoint.co.uk

British Library Cataloguing in Publication Data
A catalogue record for this title is available from
the British Library

ISBN 0 340 84836 7

First published 2002
Impression number   10 9 8 7 6 5 4 3 2 1
Year   2007 2006 2005 2004 2003 2002

Typeset by Transet Limited, Coventry, England
Printed in Great Britain for Hodder & Stoughton Educational, a division of
Hodder Headline Plc, 338 Euston Road, London NW1 3BH by Cox & Wyman,
Reading, Berks.

CONTENTS

 **PREFACE**

Since the summer of 2000, the Institute of Ideas (IoI) has organized a wide range of live debates, conferences and salons on issues of the day. The success of these events indicates a thirst for intelligent debate that goes beyond the headline or the sound-bite. The IoI was delighted to be approached by Hodder & Stoughton, with a proposal for a set of books modelled on this kind of debate. The *Debating Matters* series is the result and reflects the Institute's commitment to opening up discussions on issues which are often talked about in the public realm, but rarely interrogated outside academia, government committee or specialist milieu. Each book comprises a set of essays, which address one of four themes: law, science, society and the arts and media.

Our aim is to avoid approaching questions in too black and white a way. Instead, in each book, essayists will give voice to the various sides of the debate on contentious contemporary issues, in a readable style. Sometimes approaches will overlap, but from different perspectives and some contributors may not take a 'for or against' stance, but simply present the evidence dispassionately.

Debating Matters dwells on key issues that have emerged as concerns over the last few years, but which represent more than short-lived fads. For example, anxieties about the problem of 'designer babies', discussed in one book in this series, have risen over the past decade. But further scientific developments in reproductive technology, accompanied by a widespread cultural distrust of the implications of these developments, means the debate about 'designer babies' is set to continue. Similarly, preoccupations with the weather may hit the news at times of flooding or extreme weather conditions, but the underlying concern about global warming and the idea that man's intervention into nature is causing the world harm, addressed in another book in the *Debating Matters* series, is an enduring theme in contemporary culture.

At the heart of the series is the recognition that in today's culture, debate is too frequently sidelined. So-called political correctness has ruled out too many issues as inappropriate for debate. The oft noted 'dumbing down' of culture and education has taken its toll on intelligent and challenging public discussion. In the House of Commons, and in politics more generally, exchanges of views are downgraded in favour of consensus and arguments over matters of principle are a rarity. In our universities, current relativist orthodoxy celebrates all views as equal as though there are no arguments to win. Whatever the cause, many in academia bemoan the loss of the vibrant contestation and robust refutation of ideas in seminars, lecture halls and research papers. Trends in the media have led to more 'reality TV', than TV debates about real issues and newspapers favour the personal column rather than the extended polemical essay. All these trends and more have had a chilling effect on debate.

But for society in general, and for individuals within it, the need for a robust intellectual approach to major issues of our day is essential. The *Debating Matters* series is one contribution to encouraging contest about ideas, so vital if we are to understand the world and play a part in shaping its future. You may not agree with all the essays in the *Debating Matters* series and you may not find all your questions answered or all your intellectual curiosity sated, but we hope you will find the essays stimulating, thought provoking and a spur to carrying on the debate long after you have closed the book.

Claire Fox, Director, Institute of Ideas

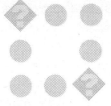

NOTES ON THE CONTRIBUTORS

Bill Durodié graduated from the Imperial College of Science and Technology in London and holds postgraduate degrees from the Institute of Education and the London School of Economics and Political Science. He is currently pursuing a doctorate at New College, Oxford, into the definition and use of the precautionary principle. He is the author of 'Plastic panics: Risk regulation in the aftermath of BSE' in *Rethinking Risk and the Precautionary Principle* (2000) and has written about science and risk for many newspapers and magazines.

Ian Gibson MP served as Dean of the School of Biological Sciences from 1991 to 1997 at the University of East Anglia (UEA). At UEA he was the head of a research team investigating various forms of cancer, including leukaemia, breast and prostate cancer. In 1997 he was elected as Member of Parliament for Norwich North, a seat he kept after the 2001 election. He is the chair of the Parliamentary Office of Science and Technology and of the Parliamentary and Scientific Committee. He is also the chair of the House of Commons Select Committee on Science and Technology.

Tony Gilland is the Science and Society Director at the Institute of Ideas. He co-directed the Institute of Ideas' and New School University's Science, Knowledge and Humanity conference in New York 2001 and the Institute of Ideas' and Royal Institution's Interrogating the Precautionary Principle conference in London

2000. He is the editor of the Institute's *What is it to be Human? –
What science can and cannot tell us*, Conversation in Print (2001).
Gilland holds a degree in philosophy, politics and economics from
University of Oxford.

Sue Mayer is the Executive Director and a founder member of
GeneWatch UK, an independent policy research group based in
Derbyshire that monitors developments in genetic technologies from
a public interest perspective. She has a background in veterinary
science, pharmacology and cell biology. She is also a Senior
Research Fellow (part-time) at the University of Sussex, a member
of the UK Government's Agriculture and Environment Biotechnology
Commission and the technical sub-committee of the Advisory
Committee on Genetic Modification.

Douglas Parr has been at Greenpeace for a number of years and has
worked on a range of issues including GM crops and food, climate
change, chemicals policy, ozone depletion and refrigeration
technology. He is currently Chief Scientific Adviser to Greenpeace
UK. He takes a special interest in the issues thrown up by the
interaction of science with society. He qualified with a doctoral
degree in physical chemistry from the University of Oxford, focusing
on atmospheric chemistry.

INTRODUCTION
Tony Gilland

Today the question of expertise, of whom we can trust to inform decisions about complex problems involving specialist knowledge, is hotly contested. On a wide variety of issues related to our health, safety and the environment we are no longer satisfied with the traditional experts, such as scientists, doctors and engineers who have trained for many years gaining specialist knowledge of a field. We are quick to question those who were once seen as authority figures on particular issues. We question their motivations and attitudes: is their advice prejudiced by some external influence such as financial reward, for example, or by uncaring attitudes towards broader social problems or the environment? We question their competence. Can they see the big picture? Do they really understand our concerns? Or are they just too arrogant, aloof and narrow in their thinking to be relied on?

And as we have questioned the role of traditional experts, new forms of expertise and types of experts have arisen. Consumer experts, environmental experts, single-issue campaigners and professional ethicists all play a greater role in official decision-making processes than at any time in the past. New areas of our lives have become subject to the influence of experts, for example in the arena of child-rearing, where parents are surrounded by a plethora of often competing advice from government, the medical profession and popular authors. So while we have challenged and questioned some of the old forms of expertise, it does not seem to be the case that we have rejected the idea of expertise altogether.

This book situates this broader debate about expertise within the controversial debates that have enveloped modern applications of science and technology. Reflecting on issues which have made the headlines in the last decade or so, it is easy to get the impression that we have been lurching from one crisis to another, particularly when it comes to the application of science and technology. The legendary scare over salmonella in eggs in 1989, which led to the resignation of the then Health Minister Edwina Currie, is now dwarfed by a plethora of additional incidents and concerns. From BSE to nuclear waste, pesticides, dioxins, genetically modified foods, child vaccines, mobile phones and global warming, there seems to be no end to the potential dangers involved in modern life. And when confronted with these apparent threats we find it difficult to know who we can trust to inform us as to the true nature of the risk and what, if anything, we should be doing about it. One thing which does seem clear, however, is that we have become highly sceptical of the advice given out by government bodies and officials as well as highly suspicious of the motivations of industry and anyone associated with it. Who, then, are we to trust?

The essays in this book have been written by people with a range of backgrounds and viewpoints. Bill Durodié researches risk and precaution at New College, Oxford. Ian Gibson MP is a scientist turned politician and is currently the chair of the House of Commons Select Committee on Science and Technology. Sue Mayer is Executive Director of GeneWatch UK and a member of the Government-appointed Agriculture and Environment Biotechnology Commission. Douglas Parr is the Chief Scientist for Greenpeace UK. Some key questions that their essays address are:

- Is scientific objectivity possible?
- How should we respond to the existence of uncertainty and risk?

- Which experts should advise on and inform decisions about new or controversial technologies and should more non-scientists be included in the process?
- How can the public's faith in scientific advice and the democratic process be improved?

OBJECTIVITY, RELATIVISM AND 'SOUND SCIENCE'

In the past the specialist knowledge of scientists was, more often than not, seen as a good thing; 'the man in the white coat' was someone who could be relied on to give us an expert opinion on an issue about which the rest of us knew very little. Due to their specialist training and research, their being part of a wider community of experts building on the insights and knowledge inherited from their forebears and the improvements to our lives that they had delivered, we assumed that scientists could provide us with more reliable information and opinions on certain issues than the 'man on the street'. And in many ways, particularly when posed in such stark terms, this is probably still the case today – no one is suggesting that we can do without scientists.

When controversy surrounds a particular technology or product with regard to its implications for human health or the environment, governments and their officials, industry representatives and scientists, argue that decisions about that technology or product should be based on 'sound science'. The idea that science has a privileged role to play in such decisions continues to be upheld to some degree, based upon the belief that it has something objective to tell us about the natural world.

Sue Mayer is highly critical of the concept of 'sound science', which, she argues, is employed in an attempt to portray science as 'impartial and objective, as the only rational basis for decision making.' According to Mayer, economic pressures upon scientists and institutional attitudes towards the environment, risk and uncertainty mean that their judgements are value laden and not objective. She argues, for example, that when scientists decide how to approach a risk assessment – what the relevant criteria are, their relative importance and the acceptability of risks – the value judgements that are involved in such decisions tend 'to be denied in their characterisation as "scientific".' For Mayer 'sound science' is shaped 'not by scientific facts but by its political, social, economic and cultural context.'

In contrast, Bill Durodié's chapter is highly critical of the increasingly popular idea that science is value laden. According to Durodié, the relativistic idea that no single view of reality is possible, only many partial ones, leaves us with a diminished model of science 'whereby separate social groups, holding unique interests, identify different aspects of reality.' He also warns of the need to recognize what is distinct about the expertise of the scientist and argues against what he sees as the corrosive impact of including non-scientific views in the scientific process: 'Far from adding to the richness of scientific enquiry, lay views necessarily tend to focus on the immediate ... Science is often counterintuitive so that without specialist training and a disciplined sense of objectivity or perspective, it becomes open to the manipulation of subjective impressions and opinions.'

Ian Gibson offers support to the seemingly opposed concerns of both Mayer and Durodié. His essay argues for the need to include non-scientific viewpoints in scientific decision making and regulatory processes, but also argues that 'it is hard to see who else to trust if these [scientific] experts are to be discounted.'

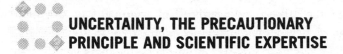

UNCERTAINTY, THE PRECAUTIONARY PRINCIPLE AND SCIENTIFIC EXPERTISE

The way in which society views the risks associated with various products, practices or technological innovations has had an important impact on the extent to which people are prepared to trust the scientific expert. Life is not, and never has been, risk free and few would deny that scientific advancement, technological innovations and economic development have all played a role in allowing us to live longer and healthier lives than previous generations. However, over the last decade or so, the idea that modern life involves in some sense a new and greater exposure to risk than in the past has become a defining feature of our times. Popular concerns about the safety of pesticides, nuclear waste, modern food production or genetic modification (GM) reflect, in part, unease about interfering in the 'natural' world. This growing unease has been informed by the idea that when we interfere with nature the outcome will be uncertain, quite possibly harmful, and if so, potentially beyond our ability to rectify and control.

Given that science and the manipulation of nature go hand in hand and that uncertainty is integral to both scientific investigation and technological innovation, it is not surprising that these concerns should have an impact on how scientific expertise is actually viewed. The centrality of this point is brought out in the essays in this book. Doug Parr argues that the announcement by government in 1996 of a possible link between bovine spongiform encephalopathy (BSE) and new-variant Creutzfeldt-Jakob disease (CJD) 'was a watershed point in collective understanding about the fallibility of science and about the possibility of long-term unknown consequences' and that 'institutional government and corporate science seemed blind to these possibilities.' The argument that scientists give too little credence to the likely occurrence of harmful unforeseen consequences of new technologies, in contrast to an increasingly sceptical public, has been growing in influence for some years. One influential Economic and Social Research Council report makes the bold claim that:

Many of the public, far from requiring a better understanding of science, are well informed about scientific advance and new technologies and highly sophisticated in their thinking on the issues. Many 'ordinary' people demonstrate a thorough grasp of issues such as uncertainty: if anything, the public are ahead of many scientists and policy advisors in their instinctive feeling for a need to act in a precautionary way.

The Politics of GM Food, October 1999

One implication of the contemporary focus on uncertainty and harmful unknowable outcomes is that, in assessing the trustworthiness and relevance of a scientist's judgement, there is a growing emphasis on the attitudes scientists adopt towards uncertainty and unforeseen consequences in addition to, or even over, their specialist knowledge and competence. Parr argues, for example, that one thing that is needed to improve public trust in scientists is an 'acknowledgement of ignorance' and that this means 'not just accepting that there are risks which aren't yet understood but that these risks are unknowable.'

In contrast, Durodié argues that there are dangers involved with the current focus on unforeseen consequences because it both 'precludes our social ability to become more resilient through development and innovation' and 'leads to little more than the creation and accentuation of unnecessary, unfounded and unassuageable fears and anxieties.' Similar concerns led the European Commissioner for Health and Consumer Protection, David Byrne, to raise the question, at a recent conference on food policy, of whether we are 'in danger of being overcome by "risk paranoia"'.

Whatever the merits of these competing arguments and approaches, contemporary concerns about uncertainty and unforeseen consequences – greatly accelerated by the BSE and CJD episode – have already had a major impact on government decision making and advisory processes and the role of the scientific expert within them. The general idea that we should be more cautious about our interactions with the natural world has been acknowledged through the widespread adoption of the 'precautionary principle' (sometimes referred to as the 'precautionary approach'). The Government's ban on the sale of beef on the bone (from 1997 to 1999); the removal from 1998 onwards of white blood cells from all blood destined for transfusion (due to the hypothetical possibility of CJD being transferred from person to person in this way); the moratorium on the growth of GM crops; speed restrictions on railways after the Hatfield rail crash; and restrictions on the use of the countryside to prevent the spread of foot and mouth disease are some of the more high profile instances where the precautionary principle has been invoked.

The essential idea behind the principle is that, in the face of uncertainty over the possible harm associated with some activity or technology, greater emphasis should be placed on providing evidence that harm will not result. In the absence of such evidence of 'no harm', the principle suggests that an activity or technology should be restricted to protect against possible (potential or theoretical) harm until evidence of safety is reliably established. This contrasts with previous attitudes, where greater emphasis was placed on there being specific evidence that harm to the environment or human health was likely to result before regulating to restrict human activities.

As it is impossible to provide 100 per cent proof that no harm will ever result from any activity, the scope for debate about how much caution should be employed at a particular point is considerable. In Britain the adoption of the precautionary approach has, in part, led the government to seek a wider range of expert advice when confronted with decisions involving uncertainty and to consult a wider range of competing groups in society. Interestingly, as Parr argues in his essay, the GM fiasco 'was not triggered by a "real" event' but rather 'the crisis was entirely a political and media creation.' For Parr, one key political issue at stake is that 'those who are expected to bear the risk are not those who create the risk' and neither are they 'likely to gain from any benefits on offer.'

Consequently, he argues, there was a 'growing body of civil society (and public mood) that was disenchanted with GM policy in the UK.'

EXPERTISE AND DEMOCRATIZING SCIENCE

Clearly, then, this debate about science and expertise revolves around more issues than those of uncertainty and our relationship to the natural world. Issues relating to industry and who benefits from new technologies and their concomitant risks and the nature of the democratic decision-making process are all at work in this discussion. This last issue has been a source of great anxiety and the Government is particularly concerned with the question of how to improve public trust in the scientific decision-making process and its own legitimacy in these matters. It is worth remembering, however, that the issue of democratic legitimacy, at a time of falling voter turnout and participation in traditional political processes, along with the loss of authority experienced by many in officialdom, are broader issues that inform and make the government's task more difficult.

In May 1999, having experienced a great deal of high-profile criticism of its policies on biotechnology (particularly agricultural biotechnology – GM crops), the government began to experiment with a more 'inclusive' and 'participatory' approach to moving forward its policies in these areas. The establishment of two new strategic advisory committees was announced: the Agriculture and Environment Biotechnology Commission (AEBC) to advise on all non-food aspects of biotechnology, such as biodiversity and other environmental issues associated with GM crops; and the Human Genetics Commission (HGC) to examine the impact of genetic technologies on humans (issues previously dealt with by the Human Genetics Advisory Committee and two smaller committees). One key stated motivation for the new structures was the view that the existing mechanisms and committees for overseeing and advising on developments in biotechnology did not 'properly reflect the broader ethical and environmental questions and views of potential stakeholders' (*Report on the Review of the Framework for Overseeing Developments in Biotechnology*, Office of Science and Technology, May 1999).

To this end, particularly in the case of the AEBC, the membership of the committees was selected to provide the government with advice from a 'wide range of interests and expert disciplines'. Thus the membership of the AEBC at its launch in June 2000 included a plant geneticist, an ecologist, a scientist specializing in nature conservation, ethicists, farmers, a consumer representative, environmentalists, a land use economist, industry representatives, a barrister in environmental law and a former presenter of the television programme *Tomorrow's World*. This was in stark contrast to the membership of the Advisory Committee on Releases to the Environment (ACRE) – which retains responsibility for advising on individual applications for new products or processes, but not the broader environmental issues associated with biotechnology previously within its remit – which then consisted of scientific experts alone. Launching the AEBC, Mo Mowlam, then Minister of the Cabinet Office, described it as 'a powerful body' that 'will be your voice in government.' She argued that the Commission's broad membership 'will ensure that the Commission looks beyond the science and beyond the regulatory framework. They need to be able to explore the issues that matter to people most.'

Such initiatives to bring a broader range of 'experts' – not just scientists – into the scientific and regulatory advisory frameworks is one aspect of what has been termed the 'democratization of science'. Ian Gibson is supportive of attempts to move away from governments' more traditional emphasis on the primacy of scientific advisory bodies and to establish new mechanisms for interaction between bodies of governance, the expert community and the general public. However, he states that 'a more pluralistic and inclusive style of governance is yet to be tested.' Sue Mayer argues for more thorough-going and radical changes to regulating and directing science and technology, which would open up decision-making processes at a much earlier stage in the process. For example, she argues that support for biotechnology has led to the 'neglect of other areas of technological innovation' such as research into organic or integrated pest management systems. Pointing to the field of nanotechnology (the manipulation of atoms and molecules of living and non-living materials) she argues it is like 'putting yourself back 15 to 20 years in the GM debate' when wider public scrutiny of the social, political and ethical considerations was absent at the early stages. Her concern is that if broader scrutiny and questioning is left to when the technology bursts on the scene 'too much will have been invested economically and intellectually to go back.'

The idea of exposing scientific research to a broader range of concerns at an earlier stage has begun to be taken up by the academic community where increasing numbers of ethical committees with lay representatives are being established to perform this role. The influential *Science and Society* report of the House of Lords Select Committee on Science and Technology, published in February 2000, endorses this approach stating that 'the Research Councils should do more to involve stakeholders and the public in the wider task of setting the priorities against which particular grants are made' but arguing that 'the scientific merit of particular research grant proposals should continue to be assessed by peer review.' However, the report recognizes that this is not a straightforward process and warns that 'to prohibit science from progressing without express public support in advance would be retrograde and repressive, and would stifle creative scientific research or drive it overseas.'

For Durodié the 'democratization of science' is highly problematic and misses a fundamental point, namely that 'while science is necessary to inform democratic decision-making processes, it is not itself democratic.' According to Durodié:

> Good decisions are not necessarily reached by consensus, especially not in science, which requires knowledge and

experience in order to be appreciated and developed. Tragically, today it seems as if the experts themselves are shying away from their responsibility to explain science to the public, preferring instead to 'include' the beliefs of their 'audience' on equal terms.

Given the importance of science and technology to all our lives, the debate about expert knowledge is clearly of great significance to the future direction of society. For Sue Mayer and Doug Parr, the current 'crisis' surrounding science and the issue of scientific expertise stems from the failure of politicians and others to recognize the intrinsically political character of scientific advice and explicitly to recognize the competing interests that are at stake. Ian Gibson, while defending the importance of scientific expertise, is highly concerned about the detrimental impact poor public confidence in scientific expertise will have on support for science and technology. From his perspective, a more inclusive process to 'prevent future and long-term clashes between all the parties involved' is the only way forward. Durodié, by way of contrast, highlights what he sees as the dangers of refusing to stand up for the ability of science to provide us with objective information and argues stridently that the 'socially inclusive approach dilutes the science, patronizes the public and allows politicians off the hook when things go wrong.'

The essays in this book aim to provide readers with a deeper understanding of some of the issues at stake and the competing arguments that lie behind the high-profile controversies that frequently surround science and technology. We hope you find them an insightful and useful stimulant to your own thinking on these issues.

Essay One

FROM GENETIC MODIFICATION TO NANOTECHNOLOGY: THE DANGERS OF 'SOUND SCIENCE'

Sue Mayer

How did we get to the situation where, when the public raises questions about the safety of genetically modified organisms (GMOs), the refrain from industry and scientists is: 'The public are ignorant, they don't understand the science.' The first genetic modification of micro-organisms took place over 30 years ago, in 1972. Plants were first modified in the mid-1980s. Plenty of time, you would have thought, for education and an informed public debate. The scientists involved in GM crop production often characterize the public outcry over GM food as having been whipped up by pressure groups and being largely non-scientific in nature (P. Dale, 'Public concerns over transgenic crops', *Genome Research*, 9 (1159-62): 1999). The implication is that these non-scientific, 'political' issues are not relevant to safety. And great effort is made to establish scientific dominion over safety assessments which, we are told, must be based on 'sound science', with experts, not the public, best placed to make the decisions. Science is cast as impartial and objective, as the only rational basis for decision making, with experts who know the science as those who must be trusted.

What is wrapped up in this seemingly unchallengeable concept of 'sound science' (who could argue for unsound science after all)? Like Levidow and Carr ('Unsound science? Trans-Atlantic regulatory disputes over GM crops', *International Journal of Biotechnology*, 2 (257–73): 2000), I believe that the phrase 'sound science' has become a rhetorical device to support a particular ideology – a political tool to control technological advance. The 'science' of those experts who support this ideology becomes mandated as 'sound science' while that of critics who take a different view is marginalized. Rather than being irrational, the public response to GMOs recognizes the subjective nature of risk assessment together with its claimed reliance on 'sound science' and is questioning its political basis.

The public questioning and rejection of GM foods has clearly disrupted their smooth introduction. But while the introduction of GMOs has presented society with a challenge for industry and governments supposed to ensure safety, the coming advent of nanotechnology (the manipulation of the atoms and molecules making up all materials, living and non-living) and the ability to recreate materials and life forms from atoms and molecules has the potential to make the current crisis look like a picnic. In this essay I examine how the risks of GM crops and foods have been handled and draw parallels to what is happening with nanotechnology.

Having followed the GM debate for over a decade I can see some problems looming. The most worrying thing being that yet again for a potentially world-altering technology, the debate is contained in the scientific and industrial communities and couched in technical and scientific terms alone when the social, cultural and ethical challenges are enormous.

CONSTRUCTING 'SOUND SCIENCE' IN THE GENETICALLY MODIFIED ORGANISMS' DEBATE

When you examine the GM debate and risk assessment process, it is clear that 'sound science' is shaped not by scientific facts but by its political, social, economic and cultural context. Although it is widely recognized that risk assessments of GMOs and other technologies involve value judgements in deciding what the relevant criteria are (the framing of the risk assessment), their relative importance and the acceptability of risks given the inevitable scientific uncertainty that exists, this tends to be denied in their characterization as 'scientific' (A. Stirling, 'Risk at a turning point?', *Journal of Risk Research*, 1 (97–110), 1998; National Research Council, *Understanding Risk: Informing Decisions in a Democratic Society*, National Academy Press, Washington DC, 1996). However, while the GM crop safety assessments have been constructed to supposed standards of 'sound science', they have had a particular and narrow focus. And while science (if it is objective knowledge only) would be expected to be interpreted identically everywhere on the globe, this is not the case.

The conventional approach to risk assessment of GM organisms focuses on the genetically modified trait and how it affects the behaviour of the organism; by making it toxic to an insect feeding on the plant, resistant to a weed killer or with higher levels of a particular nutrient, for example. However, in the USA, the claim is that the regulatory risk assessment of a GM organism does not consider the actual process of genetic modification but rather looks at the final product, because it is believed that this is what will predict its ecological effects and that this is the most scientific approach. In Europe, in contrast, the process of genetic modification is the regulatory trigger for its science-based approach although the GM trait then forms the focus of assessments. This process-based approach has been bitterly criticized as being non-scientific by many in the USA (see H. I. Miller and D. Gunary, 'Serious flaws in the horizontal approach to biotechnology risk', *Science*, 262 (1500–01): 1993). However, despite this apparent divide, a comparison of the data requirements for assessment of a GM crop shows little difference with the process of genetic modification being included (for example, the requirement to demonstrate stability of the transferred gene) on both sides of the Atlantic (S. Mayer, 'The regulation of genetically modified food', in F. Dolberg (ed.), *Encyclopedia of Life Support Systems*, Eolss Publishers Co. Ltd., 2002). There are differences in the importance attributed to different criteria, but both start from very much the same place.

The insistence on regulation being 'product based' in the USA, therefore, has more to do with the political demands of the American biotechnology industry, that genetic modification be considered no different from conventional breeding and the different way in which it is viewed in Europe. The two approaches, both claiming scientific authority, have important social consequences. In the USA, a denial of difference largely precludes labelling, while the European approach allows for this, albeit on a limited scale. International arguments about unjustifiable restrictions on trade come hard on the heels, not really because of science but because the economic benefits of GM crops might be lost if segregation is required to facilitate labelling particularly for the GM commodity crops being grown in the USA.

So 'sound science' can be affected by which side of the Atlantic you live on and the competing interests at stake. Once named, however, the way in which the assessment is then framed (what is included and excluded) involves judgements about what is considered relevant or important. In the case of GM crops and foods, the framing has been drawn very narrowly around the introduced trait. While the ecological impacts of the gene, whether it will make a crop more weedy or if the gene is transferred the effect it will have are, considered, wider issues about the intensification of agriculture, ownership of genes and compatibility with other farming systems are excluded from regulatory scrutiny. The only potential hazards included are those associated with physical impacts on the environment or human health and the yardstick of harm is the status quo.

In the case of the environment this means conventional agriculture which itself has been profoundly damaging. Yet if releasing a GM crop is calculated to do no more damage than this it is defined as safe (R. von Schmoberg, *An Appraisal of the Working in Practice of Directive 90/220/EEC on the Deliberate Release of Genetically Modified Organisms*, STOA, European Parliament, 1998). For human health, the yardstick is 'substantial equivalence' where the chemical composition of the GM food is compared, in gross terms, to a non-GM version. This has been used, not to stimulate scientific inquiry into whether GM creates unexpected changes, but rather to stifle research leaving the assumptions unexplored (E. Millstone, E. Brunner and S. Mayer, 'Beyond "substantial equivalence"', *Nature*, 401 (525–6): 1999).

Even inside this restricted framework, value judgements may vary according to the context in which they are made. In the USA, conflict about the environmental safety of GM maize containing a gene (Bt) to make it toxic to pests feeding on the crop, focused on whether resistance would emerge in target organisms compromising future use of Bt as an insecticide and strategies to avoid this. Only more recently has concern been expressed about the impact on monarch butterflies feeding on pollen from the maize. In Europe, the emphasis in the debate about environmental impacts of GM Bt maize has been on the potential for effects further up the food chain on non-target species who eat the pests poisoned by the Bt toxin (L. Levidow, and S. Carr, Unsound science? Trans-Atlantic regulatory disputes over GM crops', *International Journal of Biotechnology*, 2 (257–73): 2000).

Even inside the European Union there have been disputes between member states including debates about the boundaries of regulation, causality and acceptability which, with the changing political situation in Europe, have eventually led to the directive covering the environmental safety of releasing GMOs being revised to include some indirect effects but still excluding socio-economic issues.

This framing and its interpretation is obviously shaped by the underlying social and ethical assumptions of those involved as well as by scientific knowledge. In the case of GMOs, since the late 1980s there has been a policy commitment to biotechnology being essential for competitiveness of industry and for the economic future of Europe (Commission of the European Communities, *Promoting the Competitive Environment for the Industrial Activities based on Biotechnology within the Community*, 1991; S. Mayer and A. Stirling, 'Finding a precautionary approach to technological developments – lessons for the evaluation of GM crops', *Journal of Agricultural and Environmental Ethics*, Vol. 16, 1, 2002). The 'sound science' that the risk assessments are based on work from this unstated economic assumption both in its framing and in how uncertainty is handled. When working from a basis that there are clear benefits, uncertainties become seen in this context and their importance interpreted in that light. This brings a particular utilitarian ethical position which is not articulated as such, but as Burkhardt has argued, demands critical reflection and justification even though its proponents may firmly believe it to be the case ('Agricultural biotechnology and the future benefits argument', *Journal of Agricultural and Environmental Ethics*, 14 (135–45): 2001).

This is because technologies are not socially, politically or economically neutral but are shaped by the prevailing social, political and economic climate. Support for biotechnology has led to many social changes and neglect of other areas of technological innovation. Organic and integrated pest management systems for agriculture have suffered from lack of research investment. Patent protection has had to be extended from inventions to allow discoveries of gene sequences and their function (ill-described as this usually is) to be monopolized and fundamentally alter relationships between private and public knowledge as a result. The ability of the public to exercise choice on whatever grounds has been hampered by GM food labelling regulations designed to be based on 'sound science' which deny difference on the grounds of production and only allow it if there are conceivable health concerns from consumption. The consequences of such effects are rarely evaluated and never formally as part of the risk assessment system, yet adverse impacts there may be.

Looked at from this perspective, it is no longer tenable to argue that questioning of the basis of safety assessment is irrational or non-scientific but related to how science is being used and judged and in whose interests. In reality, the research that has been undertaken on public attitudes to GMOs or to science and risk more generally does not support the view that the public is irrational. Rather than a rejection of science, what is under attack are the institutions and processes by which decisions are made on safety and the roles that science and scientists play in them (R. Grove-White, P. Macnaghten, S. Mayer and B. Wynne, *Uncertain World: Genetically Modified Organisms, Food and Public Attitudes in Britain*, Lancaster University 1997). Fiascos like BSE have left the public all too aware that science does not and cannot provide clear-cut answers, choices have to be made and interests come into the equation. The potential for surprises remains and because GMOs are living and able to multiply, the likelihood is that any adverse impacts will be irreversible. Yet it is this potential that is denied or played down by those doing the risk assessment.

The problem for the GMO debate is that the seeds of the problem were sown back in the 1980s when the commitment to GM agriculture was made. This took place largely outside public scrutiny and inside an atmosphere of intense hype and excitement about the potential for the technology. Public anxieties when revealed in research were discounted then, as now, by experts, the industry and regulators as largely being based on ignorance. NGOs' efforts to influence were unwelcome and dismissed, rather ironically as it has turned out, as having no public support. The feeling was that the experts knew best.

There has been some recognition now that there needs to be a wider perspective on the risks and benefits of biotechnology. The UK Government has set up the Agriculture and Environment Biotechnology Commission which has the brief of looking at the wider environmental, agricultural and ethical issues arising from GM technology. Participative techniques of technology assessment and new methodologies to address the criticisms relating to the narrow framing of risk assessments are now being developed. In some ways this all seems rather late. Industry has been badly hit by the public backlash against GM foods and the public are cynical about their intentions. But what is happening with nanotechnology, and have the lessons been learnt early enough?

NANOTECHNOLOGY – THE NEW THREAT

Thinking about nanotechnology now is a little like putting yourself back 15 to 20 years in the GM debate. Drawing comparisons between the way in which genetic technologies were managed in the early days and how things are developing with nanotechnology indicates problems are in store. With nanotechnology, like genetic modification, the claim is for precision and control in shaping the future. And as with GM, how precise, under whose control and what vision of the future should be the defining questions in deciding whether we go ahead or not.

Nanotechnology (from nano meaning from 1 to 100 billionths of a meter or nanometers) involves the manipulation not of genes but the atoms and molecules making up all materials, living and non-living. The aspiration is that materials and organisms will be constructed from atoms and that waste, for example, could be transformed into useful products by reconstructing the basic atoms and molecules it contains. The technology is far from being with us; manipulating atoms and molecules is no easy matter and will rely on developing miniature production lines and nanorobots to assemble the atoms. This process will have to use the physical and chemical properties of the atoms and molecules and create the conditions where they will self-assemble. Micro-sensors for use in medicine and improved information handling for computers on a nanoscale are just two of the early applications that could be developed. While genetic modification has provided the potential to redesign life, nanotech can do not only that, but can redesign basic materials as well.

Already nanotechnology has been identified in technology foresight exercises for investment with growing research programmes in the USA, Europe and Japan. Tom Kalil, a White House economist has been quoted as saying, 'This is an area that can have a huge potential payoff that can be as significant as the development of electricity or the transistor' (R. F. Service, 'Atom-scale research gets real', *Science*, 290 (1524–31): 2000). This excitement and hype is reminiscent of the way in which genetic modification was first seized on during the 1970s and 1980s and is fuelling nanotech growth.

All this may sound very much like *Star Trek* with replicators in the corner of every room; a cornucopia of possibilities with the end of waste. But there is the potential for some very real downsides. Recently, the whistle was blown on the nanotech issue by an unlikely person, Bill Joy, co-founder and Chief Scientist of Sun Microsystems, the person who developed Java script for the internet and hardly a Luddite. Writing in an article in the magazine *Wired* ('Why the future doesn't need us', April 2000), Joy pointed to a dangerous common thread running from robotics, nanotechnology and genetic modification – that the products may be able to self-replicate. Self-replication means that problems may amplify and become out of control, which are not features of earlier technologies. With nanotech, the potential horror stories include entirely new disease causing organisms for use in biowarfare and the self-replication process getting out of control, consuming and transforming materials in the natural environment – something which has been dubbed the 'grey goo' problem.

But it is not only these physical threats that nanotech brings. It threatens to alter our working practices by replacing labour with nanorobots; damage the economies of countries supplying natural resources; and place the control of the very basic atoms of materials in the hands of those controlling and patenting the technology. It goes another step further than genetic modification in defining life and materials as amenable to mechanization. In other words, it could fundamentally alter social relations and what it means to be human – not matters a scientific risk assessment can deal with.

An analysis of the development of nanotech shows that although there are no products on the immediate horizon, publications in the scientific literature and patent applications are increasing exponentially. Commercial interest is great (P. R. Mooney, 'The ETC century. Erosion, technological transformation and corporate control in the 21st century', *Development Dialogue*, 1–2, 1999), again mirroring the development of genetic technologies where early industrial involvement shaped the development of the technology. But while commitment to the technology is growing, where is the debate about how or whether it should be applied and used? The science is dauntingly difficult to understand and simply working on such a small scale is almost impossible to comprehend. Mentioning the word 'nanotechnology' to most people would draw a complete blank. Not surprisingly, perhaps, the discussion about safety and the hopes and fears about the technology is taking place almost exclusively between scientists, with industry taking a keen interest.

The reaction to the first successful transfer of DNA between different species in 1972 came largely from the scientific community and led to the 1975 Asilomar conference in California. At this conference, scientists agreed to a voluntary moratorium on some recombinant DNA experiments until guidelines or regulations were in place. In these early days, it was only micro-organisms which were involved and the safety issue was containment and how tight that should be. Scientists were used to containing dangerous pathogens so these methods were adapted to the new demands of genetic modification. There was no intention to release a GMO and any of those in low containment which might reach the environment were considered no risk as they were unable to divide or survive. The assessment to determine containment level was narrowly focused concentrating on the GM trait and its effect, not whether the experiment was worthwhile or relevant. This approach was to become the norm even when circumstances changed in 1984, when the first GM plants were produced overturning the reliance on containment as a protection because release was a necessity to capitalize on them. This turning point did nothing to change the trait-based approach to risk assessment which had become endorsed as based on 'sound science'. Social, ethical and cultural issues were swept aside.

With nanotech a similar situation is arising. While no watershed has been passed in the development of the technology – a carbon nanotube being one of the main achievements so far – some of the scientists involved are starting to develop their own guidelines for controls. For example, the Foresight Institute, which draws many of its members from the nanotech community, has drawn up a set of principles and guidelines (www.foresight.org). These focus on limiting self-replication through containment, calling for environmental impact assessments, encouraging participation in self-regulation and limiting access to nanotech development capability to those committed to following the guidelines.

However, the Foresight Institute is also lyrical in its presentation of how nanotechnology will improve our lives:

> We will be able to expand our control of systems from the macro to the micro and beyond … Scientists envision creating machines that will be able to travel through the circulatory system, cleaning the arteries as they go; sending out troops to track down and destroy cancer cells and tumours; or repairing injured tissue at the site of the wound, even to the point of missing limbs or damaged organs … Nanotechnology is expected to touch almost every aspect of our lives, right down to the water we drink and the air we breathe.

But while there is some discussion about the dangers of nanotechnology and self-replication, biological and chemical weapons, debate about whether self-regulation will actually work and the need for any democratization of the decision-making process are notably absent. Technical solutions, such as containment for self-replicating processes, are considered adequate and the potential for accidents is played down. Moratoriums or bans on some applications as proposed by Bill Joy are the option for some of the more obviously damaging applications such as novel life forms. No word on control of the technology, alternatives and social or ethical issues.

So here, with nanotechnology, we seem to be repeating the path we took with GMOs – a discussion among the practitioners alone. Experts and 'sound science' hold sway and control the terms of the debate and the progression of the technology. But as valuable as the scientists' knowledge may be in identifying some of the potential scenarios, it is far too restricted and rarefied an atmosphere to explore all the issues. Positioning nightmare scenarios in contrast to dream worlds kept apart by only technical fixes will not reassure the public if and when the nanorobots emerge from the lab and onto the factory floor.

CONCLUSION

I do not think you need a crystal ball to predict that the public will be alarmed by the potential for accidents and mistakes as a result of nanotech. They may be excited about the possible benefits but are likely to be healthily sceptical about claims for precision and control. But at the moment the hype is gathering momentum, commitments are being made for society in which it has had no say. While nanotech of the self-replicating kind seems safely locked up in the lab and scientists happy to keep it that way at the moment, the lessons of GM tell us that the pressures may change, new justifications emerge and the prospect of more profit just too

14

difficult to resist. Although nanotech is proving technically difficult, there seems to be general agreement that the laws of physics do not preclude it, so it is just a matter of time. But what if we find in ten years' time that, as many people do now with GM crops, we want to say no thanks?

Yet hardly anyone knows about what is happening with nanotechology. There has been little scrutiny of how it may affect our lives and the implications of corporate control over matter. While scientists strive to think about the implications on their own or with industry, they are in a difficult position and vulnerable to backlash. Keeping technological trajectories and risk assessment in the domain of the expert, supported by the rhetoric of 'sound science' has failed to bring public confidence or, in the case of GMOs, to provide a stable basis from which industry can operate. Having products rejected after years of research and development is not good for business. It has also failed to deal with other complex risk issues such as BSE, ozone depletion and chemical pollution. Nanotech experts, like GM experts, are ill equipped to consider the broader social and ethical issues which become marginalized as a result. By restricting debate they also act politically to claim their dominance in the decision making and in determining social trajectories.

We need to rethink our approach to technological innovation. While nanotechnology seems distant and speculative in some respects, if it gets no public attention and influence in shaping it, if and when it bursts on the scene, the debate may be too late. Too much will have been invested economically and intellectually to go back. Alternatives will not have been developed in parallel and powerful forces will be backing its progress. As with GM food, it may become a case of forcing nanotech on an unwilling public in the name of sound science. This cannot be a productive approach, it is certainly not democratic, and we need to develop some more responsive systems which can influence technologies and related policy at an early stage. A public education programme is not the answer and neither is 'better' risk assessment. Rather we need, as Robin Grove-White and his colleagues have argued, to see technologies as social processes, not separated black boxes (R. Grove-White, P. Macnaghten and B. Wynne, *Wising Up. The Public and New Technologies*, Lancaster University, 2000).

From this new understanding we can develop new tools, with interactive dialogues between the public and other communities involved. This much broader approach should not only help us develop shared understandings of new technologies and where they should go, but also be better able to predict or cope with uncertainties and unknowns. But the options must be open, we must be able to say no – a sense that we are only fiddling with the edges of a technology and system that supports it and is supported by it will not do.

Essay Two

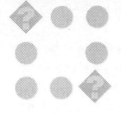

TRUST COMES FROM EXPERTISE
Bill Durodié

In 1660, what was to become the 'Royal Society of London for the Improvement of Natural Knowledge' – now the world's oldest scientific institution in continuous existence, known simply as the 'Royal Society' – was established. Its founders, wishing to emphasize the 'Experimentall Learning' that was central to the outlook of the 'new philosophy', later adopted a Latin phrase from the Roman poet Horace – the son of a freed slave – as the Society's motto: *Nullius in Verba* – 'On the Word of No One'.

This motto reflected the political mood of the time by indicating a reluctance to take on trust the pronouncement of any received authority. The challenges to the superstition and diktat of the Church and the more recently dispensed with monarch were still fresh in people's minds. Now, it was formally possible at least to aspire to replace these constraints on development through the discipline of acquired insight. Science, as well as delivering remarkable achievements, was to be a practical battering ram with which to challenge perception, prejudice and power.

But while science can effect social change, it is also driven by social change and aspiration in order to develop. Without such social progress, the direction and purpose of science can become unclear and uncertain. Today, the very authority and trustworthiness of scientific expertise is being challenged. Those who do so, hope that this will help to expose and transform the commercial and political interests they see as lying behind it, controlling its direction and priorities.

Expertise in any discipline has always proven to be uncomfortable for the ruling elite of society at any particular time. Acquired insight necessarily challenges received authority and nowhere more so than in science, which by confronting and unravelling the unknown, continuously poses new possibilities and thereby undermines existing certainties. Experts have also appeared to jealously guard the portals to their specialist areas, determining for themselves that which is good or progressive, developing their fields of knowledge by passing judgement on one another's work, and governing their own institutions, without needing to appeal to external sources of legitimacy.

This constant upheaval of the status quo, as well as the creation of a space to which they can gain only limited access, are uncomfortable phenomena for those in positions of power. Accordingly, the establishment has maintained a somewhat contradictory attitude towards expertise. At the same time as seeking to harness, draw on or legitimize themselves by association with new insights and advances, they have so sought to delimit the scope and remit of such work, holding it at arm's length and being prepared at any moment to reject it outright.

'An expert', the saying goes, 'is someone who knows more and more about less and less' or, as veteran BBC newscaster and commentator John Humphrys recently put it, someone who 'can't see the tree for the bark.' Belittling or isolating expertise in such a way can serve to justify the ignorance and actions of those who cannot access it, shape it or otherwise feel threatened by it. Alternatively, society seeks to incorporate or institutionalize expertise. Experts are increasingly found providing all manner of technical advice to policy makers at all levels of government.

Separating the provision of expert advice and the actions based on it from decision-making has also allowed policy makers and those in authority to achieve a balance between claiming credit for successful initiatives and avoiding blame for failed experiments or ideas. This has made it harder to hold those whom we elect accountable for their decisions.

But while these tensions have been in existence for quite some time, new forces have now emerged on the political landscape, which taken together threaten once and for all to demolish the authority and independence of expertise. These are the cumulative impact of philosophical relativism, the development of an exaggerated risk consciousness and recent demands that decision-making processes should be 'democratized' or made more inclusive to reflect people's 'values'. Far from leading the assault on speculation and superstition into a new millennium, it is claimed that it is the inheritors of the ideals and institutions of those who once sought to question everything, who can now no longer be trusted themselves.

RELATIVISM

Reflecting the rigid mechanistic links of cause and effect that dominated the industrial age, science was for a long time held to be based on objective, absolute and ascertainable facts. But this view of truths being revealed by pristine individuals disinterestedly recording the underlying workings of invariable natural laws does not stand up to simple scrutiny.

As Newton had already commented in his letter to Hooke of 1676, 'If I have seen further it is by standing on the shoulders of Giants' (*The Correspondence of Isaac Newton*, vol 1, 1959). In other words science comes with a history. Its advances, as well as being limited by material reality, are circumscribed by the state of the society it develops within – including its ambition and imagination.

Nevertheless, a conceptual and practical distinction between the objects of our investigations and the subjects who perform them proved to be an important one for scientific advance. Positing a singular vantage point, or a universal view of reality in which what is true for me is also taken to be true for you, helped reveal the underlying relations that shape our world.

Increasingly, over the course of the twentieth century however, philosophers of science placed greater emphasis on the uniqueness of experience. This corresponded intellectually to the tremendous changes, impasses and uncertainties they found themselves caught up in. It was becoming evident to them, that as we all bring our previous impressions and understanding to any particular situation, then the world we perceive and respond to depends in great part on our position and trajectory and thus becomes relative to our particular circumstances or past.

Accordingly, there could be no single subjective view of reality but rather many partial views. What is taken to be true then appears to be a matter of social convention or consensus. In the aftermath of war this coincided with a growing move to distinguish social from technological progress, as science itself increasingly became accused of having encouraged a narrow objectification of both people and nature. The logical extension of this has been towards the erosion of the distinction between subject and object.

Some, such as the eminent scientist and science journalist Lewis Wolpert, have suggested that such sociological considerations have little bearing upon fixed entities such as the gravitational constant or the electron mass. But relativists would argue that the very fact that society chooses to prioritize the study of dynamics or electricity over other possible avenues of enquiry is socially or culturally determined. Our values determine that which we consider worthy of investigation in the first place and hence what we discover and call science. In this view, particularistic perspectives, interests and prejudices are simply concealed behind an appeal to universality.

This approach more commonly challenges the process of scientific enquiry than the content of science. It has found a sympathetic hearing in many quarters. For instance, at a recent European Commission sponsored workshop on risk and precaution, in response to the suggestion that people's perceptions as to the impact of chemicals in the environment did not match the evidence, a Greenpeace delegate argued not against the existing science, but rather for resources to allow representatives of the public to undertake their own experiments in order to fill in what he called 'data gaps'.

This implicitly proposes a model of science whereby separate social groups, holding unique interests, identify different aspects of reality. This suggests that all we need to understand the world better is simply more empirical information from disparate sources. In a similar vein, one of the leading critics of scientific expertise, the sociologist Brian Wynne, argues for the need to 'triangulate' between the various views of academics, industrialists, politicians and the public, by which he means that in order fully to understand the contours of any issue we first need to know various positions relating to it ('*What is expertise?*', RSA fringe debate, 5 July 2001).

However science, like many other human endeavours, necessarily places more of a premium on quality than quantity. Experts do not just record or measure, they assess, infer and prioritize. Significant expertise and experience are required to make sense of experimental outcomes and decide as to their meaning. It is this qualitative judgemental mode that is most at risk of being dissolved and lost today.

Far from adding to the richness of scientific enquiry, lay views necessarily tend to focus on the immediate, rather than a mediated or more critical appreciation of available evidence. Science is often counterintuitive so that without specialist training and a disciplined sense of objectivity or perspective, it becomes open to the manipulation of subjective impressions and opinions. Incorporating spontaneous views into decision making actually corrodes the process of investigation from the inside rather than merely distorting it.

Neither does questioning alone achieve greater understanding. In many instances, it can lead to a wilful excuse for inaction, thereby precluding the ability to move away from perceptions toward a more fundamental understanding or transcendence of issues. Apart from the problems it raises for our understanding of the world and how this is organized, relativism has had a demoralizing impact on scientists, as it has the potential to strip their endeavours of all meaning and purpose. Faced by interminable questioning and being unable or unwilling to assess and prioritize, real expertise and self-confidence have been undermined. Increasingly, only empirical experience is held to matter, regardless as to whether this contains any real insight.

In the past, ruling elites sought to incorporate expert knowledge partly to justify their own prescriptive social projects. However, they could never have it all their own way, as independent and self-governing scientific bodies protected their fields of enquiry against abuse and dilution. All parties were also held in check by the need to relate to real effects and evidence. Today, as universality and the possibility of scientific objectivity have themselves been challenged, it has become far easier for the establishment to subsume knowledge to its own ends through the establishment of its own policy 'experts'. These now provide new points of reference on issues as disparate and diverse as personal health, parenting, or pollution.

Thus the Inter-governmental Panel on Climate Change (IPCC) for instance, which explores the link between CO_2 emissions from industry and climate change, while being based on the rigorous evidence of scientists, is circumscribed by the bargaining that occurs at a policy level as representatives of differing national interests seek to achieve an acceptable compromise by consensus, rather than rigorously debating the facts. Dissenting opinion or evidence has increasingly been marginalized as the new policy orthodoxy prevails. Neither does the relativization of knowledge and experience stop at the creation of new institutional arrangements. It also unremittingly criticizes the old experts and places new, and unrealizable demands on them. Doctors, for example, now regularly find themselves being challenged by their own patients who have discovered all manner of alternative cures and ailments for themselves from self-help handbooks or off the internet.

Unfortunately, it is no longer evident which of the many 'experts' is to be believed. Far from making things any clearer, the attack on expertise has made 'expertise' itself a field within which one now needs to become an expert. Those who simply rail that 'he who pays the piper, plays the tune', should note that in the absence of any clear and accepted rules of composition, the piper would never be capable of playing anything.

◉ ◉ ◆ RISK CONSCIOUSNESS

The erosion of the distinction between subject and object and the consequent elevation of subjective experience over objectivity has converged with another significant trend of recent times – the rise of risk consciousness. This phenomenon has been widely commented on by contemporary sociologists such as Ulrich Beck, whose best-selling book, *Risk Society, Towards a New Modernity* (1992), triggered much of the recent debate. Elsewhere, Beck has argued that, 'We no longer choose to take risks, we have them thrust upon us,' and further that, 'Society becomes a laboratory, but there is no one responsible for the outcomes' (*The Politics of Risk Society*, 1998).

Echoing these sentiments, Anthony Giddens, Director of the London School of Economics and Political Science, suggests that this is because in addition to the natural hazards, which we have always had to contend with, advanced societies also create new problems or 'manufactured risk'. Even if these risks can be shown not to be quite so problematic Giddens argues, they help to create the perception of a 'runaway world' (*Runaway World: How Globalization is Reshaping Our Lives*, 2000).

It is commonly assumed that the media have a significant role to play in such matters by making us more aware than previous generations of the various hazards we face. Headlines such as 'Chemical peril hidden in homes', in the usually sober *Sunday Telegraph* (1 July 2001) or the constant references to 'Frankenstein Food', which helped catalyze concerns over genetically modified organisms (GMOs) in the UK, certainly do not help to facilitate a reasoned debate. Hence, it is suggested, we are more attuned to risk and may be inclined to exaggerate it.

While there is some truth to this, what is often overlooked is the extent to which politicians, regulators and even scientists charged with ensuring safety have now adopted a more ambiguous attitude to the value of scientific expertise and evidence, as against the prejudice of public opinion. As with environmental campaigners and consumer advocates, the media then become adjuncts to this social process of the transformation of perceptions.

In his 1997 book, *Culture of Fear, Risk-taking and the Morality of Low Expectations*, sociologist Frank Furedi examined the extent to which the breakdown of many forms of social organization has left us far more isolated than before. With the decline of families, neighbourhoods, communities, religious congregations, informal associations, trade unions, political parties or other institutions to be part of, it has become far easier for our subjective impressions of the world to hold sway. This social and institutional erosion is often represented in an uncritically positive manner as a celebration of identity, choice and personal preference. Patronage and conformity have, quite rightly, been consigned to the past. But without the discipline of, and an active engagement in broader concerns, individuals have also been left incredibly isolated. When combined with the ideological impact of relativism that has eroded all attempts to hold on to some stable vision of the objective reality we encounter, it becomes easy to see how our fears, however implausible, should have become so widespread and highlighted.

We are thus in danger of replacing an old culture caricatured as being one of unthinking deference with an equally incapacitating culture of unnecessary fear. Despite being two sides of the same coin, risk is now continually emphasized over opportunity. Safety and prevention have become new organizing principles leading to a fundamental reorganization of social relations.

To preclude against the new hazards it is argued we now face, many, including pan-national bodies such as the European Commission, have called for a more precautionary approach to be taken towards development and innovation. This has necessarily impacted more upon science than most other disciplines or fields of enquiry. The drive to establish new social arrangements was brought to a head through the impact of the BSE (bovine spongiform encephalopathy) debacle in Europe, which is held to have caused vCJD (variant Creutzfeldt-Jakob disease) in humans. While vCJD does look like the human form of BSE, the link between the two has yet to be established and the episode cannot compare to other diseases, such as measles, mumps, whooping cough, smallpox and tuberculosis – which all at some stage crossed the species barrier with catastrophic effect. BSE has reinforced a fatalistic and despairing mood, providing succour to would-be reformers.

In the aftermath of the original BSE outbreak, the European Commission directorate responsible for health and consumer protection trebled its staff numbers. New scientific committees were established as well as a rapid alert system and risk assessment unit. A series of landmark documents was released, which explicitly called for the adoption of the 'precautionary principle' in all matters. This latter suggests that in the absence of definitive scientific evidence to the contrary, measures to protect the environment or human health should be taken whenever any threat of serious or irreversible damage to either may be present.

Critics have countered that, as scientific certainty is never possible, the application of the principle is a recipe for paralysis. More specifically, defining the extent of 'evidence' necessary to justify concern, as well as what 'measures' should be invoked and by whom, are considerations lending themselves to significant political, commercial and non-governmental manipulation.

Nevertheless, due to an exaggerated consciousness of risk, the precautionary principle is set to play an ever-increasing role in decision making. But as it necessitates examining worst case scenarios and also demands the inclusion of public opinion into the consideration of issues, regardless of whether a scientist would take seriously the concerns raised, it contains the potential to marginalize expertise. Thus, for instance, when in December 1999, responding to concerns raised by environmental lobbyists, the by then nervous European Commission introduced a ban against a family of organic chemicals used as softening agents in plastic, the chair of the Commission's own scientific committee was drawn to remark: 'I don't think the science is saying at all that there's an immediate risk.' Another member of his committee had already concluded that 'no matter what the scientific input', it would 'not be the decisive input anyway' (CSTEE Minutes, 7 May 1999).

The debate over the use of GMOs in food has been similarly marked by a rejection of the scientific evidence by non-governmental organizations. The potential for commercial manipulation of such issues was also exemplified when the frozen food retailer Iceland, played to people's concerns in a very public manner by withdrawing all products containing GMOs from its stock. This was a calculated strategy to differentiate its position from that of its competitors in the market-place and effectively precipitated the widespread collapse in sales of the items as other retailers felt morally pressured into following suit.

Governments themselves also seek to manipulate issues through the use of precautionary measures. For instance, rather than confronting people's concerns, the decision to establish a high regulatory hurdle for GMOs could be seen as an attempt to assuage public sentiment by accommodating to the claims of interest groups. Ironically, but predictably given the overall climate, far from reassuring, the move was seen by some as indicating that there was indeed something peculiarly hazardous about such products, which needed to be regulated accordingly. It would appear, therefore, that there is no limit to demands for, and the application of, the precautionary principle. A more insidious result, however, is the way it elevates public opinion over professional expertise and subordinates science to prejudice. Official recognition of perceptions and beliefs, then, implicitly devalue the insights acquired through detailed experimentation and detached consideration, thereby further undermining the confidence of and demoralizing scientists.

Unsurprisingly perhaps, under permanent attack and held open to constant questioning, many institutions and experts now seem to lack self-belief or even a clear vision or purpose. As a consequence policy reversals appear increasingly commonplace, thereby sending ever more confusing signals to an already sensitized public. This has led many into over-zealous reactions to events or perceived fears.

The Stewart Enquiry, into the possible hazards posed from the use of mobile phones and the siting of masts or base stations actually acted as a significant driver of public concern. Despite the enquiry proving unable to identify any evidence of harm from the phones, and even less so for the masts or base stations, the report nevertheless concluded with a general cautionary warning against excessive exposure to the technology, particularly among young people.

This was in keeping with the general mood of the time whereby it has become *de rigeur* for all enquiries to identify uncertainty and the need for caution at every opportunity. The very appointment of a government panel and the establishment of dialogue with stakeholders – whose only expertise was subjective – through citizens' juries and focus groups, gave credence to, and suggested the need to resolve, an officially recognized problem. This exaggerated sense of uncertainty, now often emerging from among the supposed scientific experts themselves, encourages procrastination among key decision makers even when there is little debate as to the science.

The decision to amend the 1990 Human Fertilization and Embryology Act to allow research into the use of stem cells from human embryos is a case in point. The process to enable experimentation in an area, which could bring tremendous life-saving and life-enhancing benefits to sufferers of a wide variety of ailments, took over two years as, conscious of the new needs to communicate and appear cautious, the Government requested still more evidence and enquiries. This led the New Labour peer, Lord Winston, a leading scientist in his own right, to describe the process as 'pathetic' and 'immoral' (cited in *Let stem cell research begin*, www.spiked-online.com, 8 March 2001).

Yet the belief that the public has a 'right to know' and should be informed whenever and wherever there is any scientific uncertainty has become one of the dominant assumptions of recent times. Accordingly, scientific experts are now spending more time brushing up on their communication skills at the expense of undertaking the fundamental research that could have led them to having something worth communicating in the first place.

Recently, it transpired that a small number of people had received blood transfusions from donors who later went on to develop vCJD. As not a single case of vCJD has ever been attributed to transmission through blood or any blood product, the National Blood Service argued that there was little benefit to imparting such knowledge.

The 'right to know' in this instance would have been little more than a right to lifelong anger and dread. Writing in relation to the growing number of young men who, having only recently been made aware of prostate cancer and despite its lack of prevalence among their age group, now routinely demand testing from their doctors, the medical writer and general practitioner, Michael Fitzpatrick, remarked: 'When clinics are swamped with the worried well, the really ill will suffer, a trend that is already apparent in many areas of the health service' (*Why "awareness" is bad for your health*, www.spiked-online.com, 30 March 2001).

Enhanced 'awareness' therefore, through a presumed 'right to know', without access to real expertise, leads to cynicism, bitterness and an exaggerated climate of concern whereby we are quite literally worrying ourselves sick. It also drives a litigious or 'compensation culture' that many deride and which as well as distorting the rational prioritization of resources elsewhere, erodes our faith and trust in each other and even ourselves. The Slovenian philosopher and psychoanalyst, Slavoj Zizek, has characterized 'endless precautions' and 'incessant procrastination' as 'the subjective position of the obsessional neurotic' (*The Sublime Object of Ideology*, 1989). Far from indicating a respectable 'fear of error', he suggests this approach 'conceals its opposite, the fear of Truth'. The psychiatrist Simon Wessely and others have similarly pointed to a correlation between enhanced concern or awareness and an increasing incidence of self-reported malaise and psychogenic illnesses (*Psychological, Social and Media Influences on the Experience of Somatic Symptoms*, conference paper, 1997).

Certainly, life has not become risk free, but what has really changed is not so much the scale of the difficulties that we face, but rather the outlook with which society perceives its challenges, both real and imagined. These problems, while different, cannot really be described as greater than those facing previous generations, neither are they uniquely insurmountable, but our collective will and imagination in seeking to resist or overcome them appears to be much weaker.

Some have argued that people are suspicious of science because they are ignorant of it. But people have a far greater appreciation of basic science today than they did a century ago. Nevertheless, previous generations were more willing to embrace scientific advance and to trust the experts. This is not because they were unaware or compliant, but because in the past there existed a greater optimism as to the possibilities and desirability of human progress. Today, it appears that each and every new technological product or process is not just scrutinized, but problematized, even, as in the case of reproductive cloning or nanotechnology, when the fears raised are well in advance of our existing understanding or capabilities and certainly before the real potential of such developments has had a chance to be realized.

To argue, as many now do, that it is the 'unforeseen consequences' of such developments which need to be guarded against is to misunderstand the meaning of the word 'unforeseen' and the spirit of scientific enquiry. Such an outlook precludes our social ability to become more resilient through development and innovation, as well as denigrating the collective human ingenuity of future generations to cope with change and difficulty. It also bequeaths a world within which new ideas are viewed with suspicion and where old ideas are continuously recycled despite their inability to deal with and explain new circumstances. And as there is always uncertainty, the constant process of communicating doubt to the public leads to little more than the creation and accentuation of unnecessary, unfounded and unassuageable fears and anxieties. The consequence will be a world within which we become far more sorry than safe.

DEMOCRATIZING SCIENCE

Many recognize that the evidence points unremittingly towards our lives being safer and longer than those of any previous generation. Nevertheless, they contend that despite clear scientific successes, the experts have failed to retain our trust. Precautionary decision making, however, is argued to act as a mechanism for inclusion, thereby restoring trust and legitimacy at a time when politics no longer engages people.

This mistrust of scientific expertise is not unique. All the old institutions, including politics and business, but also the media, doctors and even the clergy, are subject to unprecedented levels of public criticism. Many of these institutions are now seeking to restore trust through the establishment of a more active dialogue with so-called consumers and stakeholders. This is held to democratize and relegitimize decision-making by making it more participatory. The hegemony of the experts, it is held, can best be mitigated by ensuring that public 'values' become an essential part of decision-making processes at all levels. The public would then effectively be reincorporated into society as active citizens with a part to play right from the outset of any issue or investigation, rather than simply being dictated to by those whom they no longer trust.

This attempt to restore trust in science through a process of social inclusion has been encouraged by governments and accepted by scientists and industry alike. While the latter may have done so rather cynically at first, presuming that they would be restricted in their abilities to operate lest they follow the general line, it is now evident that the onslaught of relativism and risk consciousness have taken their toll. In the absence of any other direction or perceived moral purpose they now see this as the only sensible way forward and hence have wholeheartedly embraced the approach. However, such processes, which may be appropriate to the realm of politics, do not transfer so readily to that of science.

While science is necessary to inform democratic decision-making processes, it is not in itself democratic. Science is a special kind of knowledge that affords particular insight into the workings of the natural world, as well as being a necessary element to facilitate the understanding of human society. To relegate the measured and experienced judgements of scientists to being just another point of view merely reinforces the notion that science is a sectional interest, rather than a universal form of knowledge, potentially accessible by all.

The socially inclusive approach dilutes the science, patronizes the public and allows politicians off the hook when things go wrong. This is because whatever the views of ordinary people, they still only remain subjective opinions. Comparing these to the considered deliberations of experts and further holding that the latter are unable to distinguish self-interest or limitations in their understanding from objective reality or fact, both denigrates scientists and panders to the conceit of those who claim to represent the public.

Re-labelling the public's view as public 'values' and then insisting that these should be included into the policy-making process, merely aggrandizes what remain personal opinions. Opinions are open to being challenged, interrogated and altered. Calling them 'values' is an attempt to protect them from such scrutiny, which is a necessary part of scientific enquiry. Far from being open and egalitarian this is an affront to a real democracy based upon reason. Real exclusion starts when prejudice or opinion are taken to be a sound basis for decision-making.

Clearly, perception is not the same as reality. As Karl Marx reminded his generation: 'All science would be superfluous if the outward appearance and the essence of things directly coincided' (*Capital*, vol. 3, 1894). Yet the increasing use of quantitative data from polls and surveys, as well as more limited qualitative or interpretative assessments from focus groups and stakeholder forums, as a source of evidence in these debates indicates the extent to which perceptions are now pandered to and indeed being actively constructed, today. These merely reflect onto the public the prejudices of the interrogators themselves. They are then represented as the public's 'values' with an imperative to act.

Polls suggest the public have little faith in the chemical industry, for example. But this bears little relation to the improvements achieved within that sector over the last decade. Among other transformations, red-list discharges (releases of the most noxious substances), have witnessed a 95 per cent reduction, while reportable accidents to employees and contractors have halved to levels well below those found in most other industries. Research also suggests that the public is predisposed to perceiving the worst, ignoring the recommendations of scientists when these suggest little cause for concern and reacting to and extrapolating them beyond reasonable bounds at the slightest indication of possible harm. How then can including public 'values' facilitate developments, based as they are on false premises and pessimistic projections?

Nevertheless, an increasing number of businesses, concerned as to the impact of public perceptions on the sale of their products, as well as genuinely believing this to be the only way to restore trust, have indicated their willingness to accommodate and abide by precautionary decision-making as determined by committees including representatives of consumer advocates and environmental lobbies. This, they believe, will identify them as and transform them into responsible corporations operating in a responsive capacity to public concerns. But far from providing a clearer and more stable regulatory and legislative environment to operate in, the inclusion of public 'values' concedes to the prejudice, emotion and opinion of those who claim to represent the public and relegates the central role of expertise. This necessarily makes for a far more unstable, volatile and unpredictable business environment as matters are dealt with primarily at the level of perception and possibility.

As perceptions can readily be altered so, too, does this approach necessarily lend itself to facilitating far greater political and commercial manipulation into the workings of the market. It is not slmply scientists and industry that lose from the marginalization of science. For society as a whole, producers and consumers alike, a scientific approach remains the best basis for decision-making.

The use of committees, public panels and consensus groups in order to balance out differing views, or 'triangulate' between them, may appeal to woolly-minded bureaucrats in Washington, Whitehall and Brussels, but they are profoundly antithetical to science. This is not because science necessarily offers clear-cut answers to difficult problems, but rather because it uncompromisingly necessitates the judicious application of expertise in order to progress. Good decisions are not necessarily reached by consensus, especially not in science, which requires knowledge and experience in order to be appreciated and developed. Tragically, today it seems as if the experts themselves are shying away from their responsibility to explain science to the public, preferring instead to 'include' the beliefs of their 'audience' on equal terms.

The idea that better 'communication skills' or 'educating the public' lies at the root of the problem is equally misguided. Almost of necessity, education and communication campaigns into complex issues that require expert knowledge to discern their subtleties, are unspecific and overly simplistic. The recent drive to get people to cover up in the sun is a good example of this. Public anxiety is now focused on malignant melanomas associated with moles that turn cancerous. But these are relatively rare and commonly arise without exposure. Ninety per cent of skin cancers are either basal-cell or squamous-cell carcinomas for which little can be done by way of prevention. Sam Shuster, a professor of dermatology at Newcastle University has pointed to the way that melanomas are now being invented through a reclassification of malignancy designed to justify the national screening campaign (*British Medical Journal*, 1992).

In addition, the elevation of concerns – reflected at every opportunity by a government desperate to identify and appear in touch with its electorate – can backfire. The reluctance of parents to allow their children to receive the MMR (measles–mumps–rubella) vaccine is an example of this. After issues were raised by a doctor as to the possibility of a correlation between administering the vaccine and autism, the new 'right to know' industry set about broadcasting this as an equally valid view to those of the recognized experts in the field. Unsurprisingly, concerned parents – who often find themselves at the centre of such debates – stopped allowing their children to be vaccinated, a decision that could rapidly lead to a serious epidemic. Government sources then set about having to 'educate' the public as to the balance of probabilities, which had been misrepresented largely due to their own loss of nerve over other issues in the first place.

Of even greater concern has been the extent to which traditional experts themselves have, in many areas, stood by and allowed many of these developments to take place unchallenged. This paralyzing diffidence by professionals who should exercise judgement and critical thought will jeopardize more than their specialist fields. What we are witnessing is an abdication of responsibility and accountability for decision-making by the various institutions and individuals concerned.

The new elite who actively engage with and shape these processes include many with no scientific background whatsoever. Their claim to represent the interests of the disenfranchised public or the silent environment, however, is just that, a claim to a new form of authority based on their own morality or ethics. It is, of course, ironic that those who would not trust scientific expertise now have to invest their faith in a new breed of expert – the ethical expert – whose pronouncements are not at all open to any kind of experimental verification whatsoever. The arbitrary and self-appointed groups that sit on such ethics committees have no mandate to speak on behalf of anyone, let alone lecture scientists as to the merit, meaning or direction of their research. The very fact that these new moralists meet one another so frequently at conferences or on panels should alert them as to their being unrepresentative and fundamentally anti-democratic.

Little wonder then that in an age when fewer people are seeking to pursue science courses at university, more profess to be interested in or have something to say about the environment. Far from being a detailed and expert scientific understanding of nature that they bring to bear upon the discourse, what we are witnessing is the reinterpretation of environmental matters as a new morality play for our times.

CONCLUSION

The attack on scientific expertise is particularly insidious, as it takes the form of an attack on excellence *per se*. In the name of challenging elitism our prejudices are being indulged by those who seek to form a new elite. Unfortunately, this has been significantly aided and abetted by those who, having lost their own nerve in the face of uncertainty but being in positions of relative authority, have failed in their duty to the public of expanding, rather than contracting the demand for critical thought and discussion.

Far from being driven into being cautious by a risk-averse public, it is evident that this collective loss of vision and direction within governmental, non-governmental, corporate and scientific institutions increasingly inclines them into taking precautionary measures and that it is this that then feeds the broader climate of concern. Just as the initial dynamic behind the flourishing of the scientific enterprise was social change, so social change – or more particularly, its absence – has now circumscribed it too. Behind the current crisis of trust in scientific expertise lies the collapse of confidence in, or the outright rejection of, social progress. It is that broader debate that scientists will need to engage in if they are truly to regain our respect and support.

Essay Three

RESTORING TRUST
Ian Gibson

At the time of writing I am acutely aware of the distrust felt by the public in what experts tell them about transport problems, education, health, social security benefits and crime. Experts pronounce on what needs to be done and scientists and technologists are wheeled in to justify the conclusions. There is, in the UK, a scepticism about experts, politicians and the media in general, which rises above the usual British 'no can do' attitude. It has accelerated through the exposure of the public to issues such as BSE, GM crops, global warming, depleted uranium, stem cells and foot and mouth disease. We should not, however, exaggerate the permanency of public scepticism towards expertise given that polls also indicate recognition of the importance of expertise. So what does current evidence tell us about public attitudes towards science and scientists? How have government and policy makers responded? And what are the issues that matter for restoring the public's faith in scientific expertise?

SCIENTISTS AND THE PUBLIC

A far-reaching report involving MORI polling makes the point that the public has clear opinions about when and why it trusts science and scientists (R. M. Worcester, *Science and Society: What Scientists and the Public Can Learn from Each Other*, 2000). One of the central issues is the question of 'who benefits?' Some scientific developments are recognized as beneficial for everyone, primarily those that bring medical benefits and this improves people's trust in those developments. This contrasts sharply with far more sceptical public attitudes towards genetically modified food where little or no benefits are believed to accrue to the wider public. Trust is also improved when science is perceived to be well regulated. MORI found that respondents lacked a general knowledge of how science is regulated but nevertheless believe that there is little regulation. This perception heightens public hostility towards new technologies and those involved with producing and regulating them. MORI found, however, that scepticism and mistrust of government and business is not always accompanied by a distrust of scientists themselves. Scientists appear to be most trusted when either working for a university or an environmental non-governmental organization (NGO) and least trusted when working for government or industry. Despite mistrust, the MORI poll found science is seen to be important to people. Nevertheless, scientists clearly have much to do to convince the public that *they* are concerned about their work benefiting the wider public. What do we know about scientists' attitudes towards their relationship with the public?

Another poll carried out by MORI, commissioned by the Wellcome Trust and supported by the Office of Science and Technology (OST) tackled the questions of whether scientists consider themselves to be the people most responsible for and best equipped to communicate their scientific research and its implications to the public; what benefits and barriers scientists see to a greater public understanding of science; and what needs to change for scientists to take a greater role in science communication (*Science and the Public – A Review of Science Communication and the Public Attitudes to Science in Britain*, 2000). The key findings were:

- There is a large gap between the way in which scientists perceive themselves and the way they think the public perceives scientists. Scientists have a far more favourable image of themselves than they think the public has of them.

- Most scientists can see benefits to the non-specialist public having a greater understanding of science, but most can see barriers too. Three in four scientists regard a lack of public knowledge, education and/or interest in science as a barrier. A third also perceive the media as a barrier to the public's understanding of science.

- The main sources of information which scientists think the public use to find out about scientific research and its implications are national newspapers and television. Scientists think that the public primarily trusts the media and those working for charities and campaigning groups to provide accurate information about scientific developments and the associated risks and benefits. They themselves are most inclined to trust those working in scientific circles to provide such accurate information.

- The vast majority of scientists believe it is their duty to communicate their research and its social and ethical implications to policy makers and to the non-specialist public. A clear majority also think that scientists should report on any social and ethical implications of their work when publishing their research findings. However, many scientists feel constrained by the day-to-day requirements of their job which they believe leaves them with too little time to communicate with the wider public or even to carry out their research.

- Scientists mention a variety of groups as being the most important with which to communicate, indicating a broad potential audience with whom they could communicate. Most scientists feel that they themselves should have the main responsibility for communicating the social and ethical implications of scientific research to the non-specialist public. However, fewer feel that scientists are the people best equipped to do this. Just over half of scientists have participated in one or more of 15 given forms of communications activity in the last year. Participation is related to scientists' skills and confidence: those who feel equipped to communicate the scientific facts and implications of their research, and scientists who have received training, are more likely to have participated. Similarly, scientists who teach, as well as conduct research, and who therefore have experience of communicating to non-specialists, are more likely to have communicated in the past year.

- Three-quarters of scientists feel equipped to communicate the scientific facts of their research, although only one in five feels very well equipped. Confidence declines when scientists are asked how they feel about communicating the social and ethical implications of their research. Among those whose work has social and ethical implications, 62 per cent feel equipped and one in ten feels very well equipped.

- The overwhelming majority of scientists have not been trained to liaise with the media or to communicate with the non-specialist public. Most scientists are aware that their institution or department provides a range of communication services. In contrast, relatively few scientists are aware of the communications services provided by other organizations, such as those who fund scientific research.

- Scientists mention a wide variety of stimuli to improve communications. Incentives from funding authorities to encourage time spent on science communication are mentioned most frequently, followed by training in dealing with the media.

Such surveys indicate that there are important difficulties in the relationship between scientists and the public. Scientists perceive themselves to be both misunderstood and mistrusted. While they believe they have a great deal of responsibility for communicating their findings to the public, many also believe that they do not have the time or the specialist skills to do so.

The fraught relationship between scientists and the public has been heightened during crisis situations and public debates over issues such as GM crops, foot and mouth disease and animal experiments. All too often scientists have been taken by surprise by the levels of public concern generated by these issues and have frequently given out contradictory messages. Scientists have also suffered from the perception of being in the pockets of industry and government which has been accentuated by the extent to which research has become more and more dependent on charitable and industrial finance. However, what is meant by independence and how it can be proved is far from obvious.

The popular perception of scientists, among young and old alike, is that they are clever but 'not of this world'. The 'Dr Who' image – the eccentric – is still the dominant one of the scientist. In a recent debate on stem cell research in the House of Commons, the shadow spokesperson, talking of science and scientists, exemplified this view: 'I suggest that the age of deference to scientists is over. It will seem to many people that science has failed us in many spheres, and the fact that there is a lack of proper control over scientists is – with, as always, the benefit of hindsight – obvious.' He went on to argue: 'I think that it would be very dangerous for scientists, if science were constantly pushing ahead of where the body of society is comfortable and happy to be. It is incumbent on those who seek to make such changes to carry public opinion with them' (*Hansard Parliamentary Record*, 17 November 2000).

It is unlikely that the days of complete subjugation to the scientific expert will ever return. They will have to take time, and trouble, to explain what they are attempting, how it will affect the public and the environment and, indeed, what benefits might arise.

SCIENTIFIC EXPERTISE

As professionals, scientists like to set their own standards and to monitor them through the process of peer review. This means that only other scientists who are also experts in the field they are reviewing can effectively comment on and judge the work of any scientist. Problems are, for the most part, resolved in-house unless journalists or whistleblowers raise concerns, for example, highlighting cases of bias through vested interests. However, some forms of external influence over the work of scientists have begun to emerge. Notably an array of ethics committees has been established to monitor the work of scientists and the ethical implications of scientific developments. These committees are required to include lay members as well as scientists. For example, the Human Fertilization and Embryology Committee, which monitors research using sperm, eggs and human embryos, is required to have a 50 per cent lay membership.

Some have objected to the idea that the scientific advisory system should address ethical or value-based subjects, although this is very much out of kilter with government thinking. In response to such objections, Sir Robert May, the former Chief Scientific Advisor to the Government, answered by saying: 'There is a need to involve ... some experts from other, not necessarily scientific disciplines to ensure that evidence is subjected to a sufficiently questioning review from a wide ranging set of viewpoints.'

Public appreciation of the role of scientific expertise is also complicated by the fact that there is invariably disagreement within the expert field. The question of whether we should vaccinate farm animals or not was a major focus of debate among scientific advisors during the foot and mouth outbreak. The need to address problems where there is scientific uncertainty and where there is a range of scientific opinion is the norm. This has significant implications for sensitive areas of public policy.

Food safety is obviously an important area where trust in scientific expertise has been undermined. Results of a survey recently presented at the Royal Society in London show that 'trust in government and its agencies, as well as business, appears patchy in relation to food safety issues.' Interestingly, however, the newly created Food Standards Agency – which has emphasized the importance of openness and transparency in its decision making – performed incredibly well. In a rating from 1 to 5 (1 being least trusted and 5 the most trusted), the Food Standards Agency scored close to 4 while ministers and government departments ranked on the lower side of the scale (just over 2). The survey also found that 'openness may be a necessary but is certainly not a sufficient condition for achieving trust.' The trust being placed on an individual or organization is related to them exhibiting (or being perceived to exhibit) a number of attributes:

- independence (from stakeholder or political influence)
- expertise (in the problem domain)
- perceived public interest (above all else)
- consistency of position
- actions congruent with words
- adequate means (for example, resources) to meet objectives.

The sooner the public develops its own expertise not based on degrees, knighthoods and so on, so much the better. Honest questions raised by intelligent members of the public with the 'experts' should now become the order of the day. Expertise comes in many forms and does not always rest with the scientist.

TACKLING THE PROBLEM: THE UK GOVERNMENT'S APPROACH

The 1999 review by the Cabinet Office and the Office of Science and Technology of the advisory and regulatory framework for biotechnology recognized that current arrangements are 'too fragmented, are difficult for the outsider to understand, lack transparency, do not clearly take on board the views of all potential stakeholders and broader ethical and environmental considerations, and are insufficiently flexible to respond to the fast moving nature of biotechnology development' (*The Advisory and Regulatory Framework for Biotechnology; Report from the Government's Review*, 1999). Identification of this problem of governance led to a policy solution: an advisory and regulatory framework with three commissions adopting a strategic oversight in the areas of health, agriculture and food, supported by 14 specialist regulatory committees.

The political challenge of this new framework is to manage the paradox between protection of the public and facilitation of industrial progress. This will require a dialogue between regulators, political channels (ministers), the industry, the interest groups and the consumers. Scientific expertise can no longer act as the authoritative source for risk regulation and it must become a constituent element in the political discourse.

This, of course, raises questions of the membership of the advisory and regulatory frameworks, commissions and committees. A more pluralistic and inclusive style of governance is yet to be tested. Representations of interest, scientific knowledge, industrial influence, nominations of individuals as experts or non-experts make for a series of complex relationships. When it comes to human genetics and health, with its vast economic potential, the media and public become increasingly interested. The role of pharmaceutical companies and the closed world of medicine regulation will require new beginnings and examination. Government still remains ambiguous on the question of the number of experts and lay members required to be on advisory committees, reflecting the paradox of trying to serve industry and the public.

A recent suggestion by Sir John Sulston, famous for his involvement with the human genome project, is for a scientific equivalent of the Hippocratic Oath. The idea was that a commitment from scientists to promise to 'cause no harm and to be wholly truthful in their public pronouncements' would help to allay distrust of scientists and to protect them from discrimination by employers who might prefer them to be economical with the truth. This is unlikely to succeed in the competitive world of grant allocation, short-term contracts, poor salaries, the increasing role of industry in academic financing (as demonstrated in the USA in particular) and the encouragement of industrial type innovations within pioneering, 'blue-sky' scientific research. The study carried out by the House of Commons select committee on science and technology on the British biotechnology industry (*The Regulation of the Biotechnology Industry*, 1999) illustrated the conflicts between commercial interests and basic scientific pursuits like clinical trials. This study led to a self-regulatory code of practice being adopted by industry and science but it took a whistleblower and a select committee report to develop it.

One issue becomes clear: that new mechanisms for interaction between bodies of governance, the 'expert' community and the general public (active or passive consumers) are needed. These will, in some instances, prevent future and long-term clashes between all the parties involved. Recent examples, such as the Hampshire waste management project, point in this direction. Here, a community involvement programme delivered a higher level of consensus on the possible waste management options to be implemented (*Open Channels: Public Dialogue in Science and Technology*, report by the parliamentary office of science and technology, March 2000). There is indeed a need for public dialogue, but unless this is characterized by its fairness, appropriate methodology and timing, as well as by its definition of participation (who is to take part in the dialogue and for what purpose), it will not deliver. Unless new forms of communication are perceived as legitimate they will only increase the level of apathy in the population. I now turn to consider some other important dimensions of the problem of lack of trust in scientific expertise.

THE PRECAUTIONARY PRINCIPLE

In areas where there are reasonable grounds for concern about a new technology and its risks to human and animal health or the environment, it is fashionable now to invoke the precautionary principle. In a nutshell, the precautionary principle stands for a policy approach that recommends cautious reaction in cases where there is scientific uncertainty about possible harmful effects of a new technology. It advises that in order to prevent possible harm, preventive steps need to be taken, even if possible harm is 'theoretical', that is, cannot be proven. This often occurs when the scientific evidence and data available are insufficient to allow a detailed risk assessment to be made. If the available scientific data point to some adverse effect and sufficient risk is perceived, this can trigger usage of the principle. Opposition to the use of the principle comes where it is seen that it might prohibit manufacture, distribution or some other part of the industrial process. This can lead to tensions between different groups of scientists or between individual scientists and their organizations, if their respective risk assessments differ and this leads to the application of the precautionary principle by policy makers.

A recent report on mobile phones has called for a precautionary approach to their use because of the lack of scientific evidence one way or the other as to their safety in terms of human health (*Mobile Phones and Health*, report from the Independent Expert Group on Mobile Phones, Chairman Sir William Stewart, 2000). The same approach has been taken in the context of banning beef on the bone, advice to dentists to avoid mercury amalgam fillings and the single use of surgical instruments for tonsillectomy. All these have involved precautionary responses to 'theoretical risks'.

The principle, with strong support of governments and advocacy groups, will become a major component of risk assessment and decision making in the face of a lack of scientific understanding. It should be seen as an opportunity and not a threat. Whether it will help restore public confidence in science is dubious since political decisions will rarely be based on the expertise of scientists alone and not while innovation and entrepreneurship still dominate in the current political environment sweeping aside all principled arguments before them.

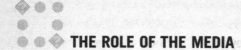

THE ROLE OF THE MEDIA

A recurring theme in the expertise argument is how the media handle some of the problems I have mentioned. This reached its zenith in the debate on GM food. The House of Commons science and technology select committee, when looking at the fashioning of public opinion in the light of expert advice, commented on the politicization of science issues such as GM food and mobile phones and the volatility of public opinion. The committee stated that measures are needed to stabilize the science-media relationship and to encourage a rationale debate without which GM technology and its potential benefits may be permanently lost to the UK. The training of students and staff in handling the media and the development of a code of practice for media coverage of scientific matters were major recommendations. In this way it was felt that scientific discoveries and new technologies would be accepted by the public. Reference to the 'bad' media coverage of scientific issues as witnessed by references to 'Frankenstein foods' and pictures of a green-faced Tony Blair is often blamed on the political journalists taking over the story to spice it up and to 'lay it on the Government.'

Science writers in both the tabloid press and other papers tend to be less likely to capture the headlines than their colleagues. The attempts by journalists to go for the big headline, contrast with the more balanced assessment needed to examine, for example, the reasons to vaccinate, or not, cattle herds before they were moved during the foot and mouth debate. The media, both the press and television, are to be congratulated for raising and indeed exposing problems in the face of a lack of scientific understanding and often in the face of lethargy or downright unwillingness to hear from the scientific community. However, when the press turns to scientific experts for advice and answers this can very rarely be given in black and white terms. This is annoying for the media, so they are tempted to decide on the answer required for the headline.

All this makes the public even more sceptical as they read and watch a story unfold. They often pick up the contrasting views of stakeholders competing for influence, be they regulators, scientists, politicians, the industry, interest groups or, of course, the public. Within each one of these groups there are of course further internal tensions between the factions. There is rarely one voice to speak for science.

DO WE NEED EXPERTS?

The combination of scientific peer review and scrutiny with the economic self-interest of industry are sufficient, some believe, to maintain public confidence. The belief is that remedial action will be taken in this arena of joint scientific and industrial self-regulation. Where the state has to decide, it is ministers and government agencies that have the ultimate status. A particular expert committee may even be set up with statutory powers to advise ministers on exercising their power on a specific piece of legislation. Most regulatory and advisory committees, however, do not have such powers. Many committees of experts are then merely ad hoc. Neither is there only one minister involved in making the decision. Indeed, in the biotechnology governance arena with the new regulatory framework of 1999 there were five ministers involved.

The current government seeks to regulate public opinion by winning hearts and minds. This accentuates the belief that good science or, indeed, bad, will win the public if you can manipulate it through the media. Thus expertise can become a question of competent communication through the media. Lobby groups and vigilant reports have countered such an approach and this has resulted in people being driven from quiescence and turned into angry shoppers. Monsanto, Novartis and other biotechnology companies have felt the backlash of this lobby. There is a collision course with the scientists caught in the middle, manipulated by government and industry and distrusted by the public.

In a recent statement to the biotechnology industry, Tony Blair said: 'Human progress is driven by both science and moral judgement. Scientific innovation has been the motor and judgement the driver.' The Prime Minister went on to acknowledge that 'there is as yet little wider public understanding of the revolutionary potential of biotechnology.' It may be he has yet to be convinced of the essential need for public support and more so of the examination of the methods needed to bring this about. It is one thing to support industry, another to convince the public. A document published by the Office of Science and Technology (OST) in 1997 calls for information from a variety of sources to enable a full appraisal of a problem (The use of scientific advice in policy making). It includes the OST's own research, international bodies, expert and advisory systems, the research community, industry and an academic survey.

Issues emerge, however, with little or no prior warning. The question which then arises is whether a department has the capacity to recognize the implications of and to react quickly and efficiently to emerging crises. OST maintains it keeps trans-departmental issues under scrutiny and control. However, the events involving the then Ministry for Agriculture, Fisheries and Food (MAFF) in the recent foot and mouth outbreak and their apparent dilatory response of how to engage science culminated in the Food Standards Agency (FSA) and the Chief Scientific Advisor taking over.

A Department of Trade and Industry (DTI) document of 1998 argued that they should ensure that, in future, government bodies:

- draw on a sufficiently wide range of the best expert sources both within and outside government
- take independent advice of the right calibre ensuring impartiality of advice is not called into question
- ensure research is undertaken when appropriate
- involve experts from outside the UK (Europe and other international sources)
- involve experts from other, not necessarily scientific, disciplines.

The European Community may have the competence to help in decision making. There ought to be interaction at the European Community level, with exchange visits of scientists. A key factor in this process, however, must be to ensure that civil servants have some scientific training to ensure they understand data assimilation and interpretation. Without this the whole process could fail. There is no use delivering inaccurate information to the process.

CONCLUSION

Clearly, there is little trust and respect for scientific experts at the current time. It is hard to see who else to trust if these experts are to be discounted. The inclusive process of involving experts from various sources including the Government and the public has yet to be evolved to ensure public confidence in science and acceptance of new techniques or technology. Without this support there are likely to be major problems for the Government and the regulatory system. The sooner a wide variety of experts engage in the political process and with the public, the better. A scientific expert 'community' needs to be formed which responds to these problems or they will just return and drive down the support for science and technology to the detriment of society.

Essay Four

KNOWING ABOUT IGNORANCE
Douglas Parr

How can today's community of scientists regain the trust that was shown to their forebears? I will argue that they cannot without substantial changes in science's relationship with society. This essay will look at the origins of the current mistrust; the social factors that underlie its most recent expression, the hidden values in scientific assessments, the supposed 'remedies' that are being suggested to this lack of trust and, finally, at a genuine way forward. I am not going to talk about 'image' as I believe this to be secondary. Having said that, pictures of Italian professors planning to go ahead with human reproductive cloning while it remains deeply antithetical to so many people is unlikely to enhance the profession's standing.

To begin, let's look briefly at the origin of the mistrust in science and the issues that have characterized it. In recent years this has been crystallized by bovine spongiform encephalopathy (BSE) and its associated human form, nvCJD. The denial of possible risk, systematically and over a period of years by British ministers, was the prelude to a collapse in British beef sales at home and abroad. It has coloured attitudes to beef for years after the announcement in 1996 that there was, in fact, a possible link between BSE and nvCJD. As researchers from the University of Lancaster have argued, BSE was a watershed point in collective understanding about the fallibility of science and about the possibility of long-term unknown consequences. Institutional government and corporate science seemed blind to these possibilities. Importantly, the term used is 'watershed' not 'revelation'. BSE emerged in a society where many people already had misgivings about scientific understandings in relation to issues such as doses of low-level radiation, the impact of pesticides, environmental impact of synthetic chemicals and, from longer ago, asbestos. All are about the imposition of risks on members of the lay public. In each case, denials of any undesirable effects were systematically undermined as new evidence became available. This situation arises in part from common assumptions in the scientific assessment processes which are explored later in this essay.

However, more subtly, there are features to these conflicts over environmental and health risks that go wider than the boundaries imposed by available scientific evidence. In each case those who are expected to bear the risk are not those who create the risk. Neither are the 'risk bearers' the ones likely to gain from any benefits on offer, which again go to the risk creators. Thus there is inequality: an intensely political problem that with current approaches can only be made tractable by saying that the risk is zero and that therefore there is no political question to answer. But that cannot be accurate where new innovations inevitably carry with them some level of ignorance about their effects. Experience suggests that there will be 'surprises'; indeed surprises are a normal and important part of the scientific discovery process. The public is effectively being asked to suspend their disbelief that these surprises will not occur when the science leaves the laboratory and enters the messy reality of the everyday world.

Further, with nuclear radiation, the effects of chemicals and BSE, the risks have particular characteristics: they are poorly understood, invisible and have impacts a long way from the source of the problem in time or place (or both). It has long been known by social scientists that people are more averse to these kinds of risks. They are also those for which science has most difficulty in establishing cause and effect relationships and so they are most likely to spring 'surprises'.

THE GENETICALLY MODIFIED FOOD FARRAGO

In the light of these surprises, we could say that there had been some 'social preconditioning' to the most recent uproar over a new application of scientific knowledge, genetically modified (GM) food. Seen against this backdrop – the absence of a political process for risk, characteristics that are likely to make people risk averse, a history of risk denial being found out – the farrago over GM was almost inevitable. Perhaps what was surprising was not that it happened but that its media expression was not triggered by a 'real' event – the crisis was entirely a political and media creation. As has been documented for the Parliamentary Office of Science and Technology (POST) in its scrutiny of the intense media coverage of February 1999, it took place against a backdrop of a growing body of civil society (and public mood) that was disenchanted with GM policy in the UK – as represented by, for example, the enormous growth in the number of health and civil society groups signing up to the 'five-year freeze' campaign (a campaign calling for a five-year moratorium on the introduction of GM crops and foods until further tests have been carried out to establish their safety for human health and the environment, see www.fiveyearfreeze.org).

However, the GM uproar was most fundamentally driven by a deep mismatch between government policy and public mood in respect of the approaches to, and understandings of, scientific uncertainty and ignorance. As Professor Brian Wynne, having conducted research into public attitudes to GM across five EU nations, argues:

The most general finding was of a profound dislocation between, on the one hand, understandings of uncertainty which scientific and policy institutions expressed and reflected in public discourse and decision and, on the other hand, understandings of uncertainty which informed typical lay public concerns about the management of the GMOs issue. The institutional understanding was based on the explicit exclusion of ignorance – only known and observable uncertainties were given standing in policy consideration; and these were assumed to be fully tractable to further research where necessary... Against this typical public thinking was ... that people did not expect certainty – indeed quite the opposite they took uncertainty for granted; but the uncertainty that they took as salient was *ignorance*. Their concern was not just that ignorance about all the possible consequences existed – they took such lack of control (including lack of intellectual control in the form of prediction) for granted based in hard past empirical experience. It was that institutions were effectively denying this state of affairs by focussing attention only on *known* uncertainties... There was lack of confidence in science because what they recognised as the predicament of ignorance was being effectively denied, and responsibility for handling it also denied, by scientific and policy institutions.

[emphasis in original]

(Paper presented at the Biotechnology and Global Governance conference, Kennedy School of Government, Harvard University, April 2001)

These concerns extend way beyond the borders of the UK. Sadly, the same mistakes continue to be made in the UK with GM crops policy. In the face of public furore the response was 'a bit more science will sort it out – let's start the farm-scale trials of GM crops.' Never mind what worries people, never mind that public concern lies elsewhere.

Stick to the unshiftable attachment to science as a means of resolving political issues. It is astonishing how many people, even quite close to the GM debate, do not realize that the farm-scale trials (designed to compare the environmental impacts of farming GM crops with the farming of conventional crops) are not about the process of genetic modification itself but are about the agricultural practices and herbicide use associated with the farming of GM crops. The ignorance about the wider issues of GM technology is ignored and responsibility for handling it denied.

VALUE JUDGEMENTS

What is intriguing about the GM crisis is how science and its institutional representatives became so closely involved with the promotion of the technology. The prized 'objectivity' of scientific endeavour was closely allied with a genetically modified organism (GMO) approval system that clearly had value judgements embedded in it. Examples of the issues around which the regulatory framework and scientific assessment process have come to a value-laden view are:

- *That, as outlined in the quote we have just seen from Brian Wynne, the attitude to risk acceptability commonly found in scientific assessment processes (denial of ignorance) shall prevail over public reactions to uncontrollable, poorly understood, inequitable, intergenerational and potentially catastrophic risks.* But these public reactions are not, as often characterized, 'irrational' but represent perfectly rational judgements based on experience and reflection; that the possibility of significant ignorance is being ignored.

- *That GMOs are a good thing and should be approved unless there is specific evidence for a specific problem with a specific GM release.* This fails to acknowledge the possibility that there may be unidentified hazards and the fact that the risk makers are not the risk takers. This lack of fairness raises the need for a justification of such risks both to an individual's health and to his or her environment.

- *That nature is being respected and unacceptable boundaries are not being crossed.* However the use of genetic engineering raises for many people questions about humankind's relationship to and respect for nature – questions which arise both from religious and secular concerns. In terms of crop plants these 'ethical' issues have no point of entry in the regulatory process.

- *That further 'interference' with food is not objectionable.* In contrast, a strongly felt desire in both opinion polls and consumer behaviour is in the direction of organic produce or more generally less chemical-intensive, less industrialized food rather than more technical processes in food production.

These value judgements shaping the policy process are clearly ones which Greenpeace, for one, would not share. In practice, I suspect that most members of the lay public would not share them either. The Royal Commission on Environmental Pollution pointed out in 1998 that: 'Values are an essential element in decisions about environmental policies and standards. People's environmental and social values are the outcome of informed reflection and debate.'

It is more than possible that the 'informed reflection and debate' is a better framework than the hidden, embedded assumptions in the scientific/regulatory system. Similar sentiments were expressed in the House of Lords science and technology committee report on sience and society in 2000. These values are where campaigns like the ones conducted by Greenpeace gain their influence. Unless the values to which the campaigns speak are in resonance with the wider public mood they will fail. Equally, scientific consensus is not enough to carry forward political change. Climate change is a classic example. While there is undoubtedly more that could be done to reduce the scientific uncertainties over climate change, the overwhelming majority of evidence points clearly to its being a major threat. Sometimes these value-based objections can (or could) be dealt with by reframing the issue; in the case of climate change, even at the time of the fuel protests in September 2000, polling suggests that there was significant support for fuel taxes if put into alternatives such as public transport and to allow diversification into green fuels like biodiesel. However, in the case of GMOs the values being transgressed would seem to be more fundamental.

ERRONEOUS ASSUMPTIONS AND JUDGEMENTS

We shall return to the GM debate later, to examine establishment reaction to the need to 'regain trust'. But for now we should look at some of the judgements that have affected technical evaluations on issues past and present to show that that GM is not an isolated case. Here the focus is not on values but on the inherent judgements that have been mistaken, most definitely not just that more and better science was needed. The point in going through this is to show that the attitude towards uncertainty within science (or, at least, science as deployed in policy processes) is long standing and pervasive, so that the public has good reason to be suspicious of the weaknesses of scientific assessment. In other words, the lack of trust currently being displayed towards the way science is used in policy has a good evidential basis.

Policy on chemical and radioactive discharges has long been based on the assumptions of:

- *assimilative capacity*, that is, that the environment has the ability to absorb and deal with a limited amount of particular chemicals
- *dilute* and *disperse*, in other words that if something nasty is diluted enough it will stop being nasty enough to have any impact.

This has for a long time underpinned, for example, the Environment Agency approach to discharges or the Pesticides Safety Directorate approach to pesticide regulation, on a chemical-by-chemical basis. These assumptions are badly flawed for at least three reasons:

SYNERGISTIC EFFECTS AND COMPLEXITY

Chemicals rarely act singly. Usually ecosystems, particularly aquatic ones, will be subjected to a variety of stresses and multiple chemical impacts. Any risk assessment system which looks at each chemical assault separately from others will not really gain a proper assessment of environmental or health impact. Only rarely, for example for dioxin toxicity, are there models routinely applied for effects of more than one chemical at a time. Even in this case, combined impact is accounted for through summation of individual effects; there are plenty of instances where such an additive model is manifestly inadequate to known situations. For example, toxicity of dioxins in higher mammals has been estimated from body burdens measured from fat sampling. However, dioxins are never found alone but in combination with other persistent organic pollutants, for example PCBs (Polychlorinated Biphenyls), whose toxicity is poorly characterized. The impact of synergistic effects is compounded by the likely complexity of chemical exposure as a result of discharges from industrial plants where compounds unknown to the operators are produced and chemical reactions in the discharge pipes produce further chemicals. These effects have long been known about but only recently have regulators looked to change the basis of regulation to look at overall biological assault. For example, the Environment Agency is trying to look at combined effects by changing to biological indicators – although for more detailed reasons their approach is not a panacea for addressing issues of complex chemical exposure. Further, the effect of complex exposure may be widespread – evidence suggests that the combination of chemicals associated with the drilling for and production of oil in the North Sea can have effects up to 6 km away from the installation.

NEW PATHWAYS

A founding assumption of the risk assessment process is to be able to establish the hazard, then to establish the probability of the hazard to give an estimate of the risk. However, risk assessment fails to deliver if the hazards are unanticipated or result from combinations of circumstances that were not anticipated. Perhaps the best example of this is the emergence of endocrine disruption as an effect of anthropogenic chemicals. Tri-butyl tin or TBT and its effects on whelks, where all the whelks affected by tiny concentrations (tens of parts per trillion or thereabouts) became masculinized, is the touchstone of this emergent impact in the early 1980s. Endocrine disruption may be a very widespread phenomenon among synthetic chemicals. Equally, radioactive isotopes in the environment have repeatedly been found to behave in unanticipated ways. Examples include the ability of both seaweed and lobsters to bioaccumulate technetium-99 to much higher levels than those found in seawater. (That is, for these creatures to make the technetium-99 much more concentrated in their tissues. Technetium-99 is an artificial radioactive element mostly coming from discharges from nuclear installations.) Pigeons also accumulate radioactive material in ways entirely contrary to existing models of pathways of exposure for people and animals. This was also true of the long-standing restrictions of sheep movements from farms in North Wales and Cumbria when initial assurances had been that the contamination from Chernobyl would last only a few days or weeks and yet, in fact, lasted over a decade. The classic of all these new pathways is the discovery of a protein that was able to make copies of itself without intervening DNA or RNA machinery of the cell – the causative prion agent in BSE.

LIMITATIONS ON AVAILABILITY OF DATA OR RELATING CAUSE AND EFFECT

It is a truism to say that both chemical and radioactive discharge to the environment have occurred without full knowledge of their effects. This is despite, for example, inadequate knowledge even about basic hazards with respect to chemicals. For example, the European Environment Agency estimated that only one quarter of chemicals on the market have even basic hazard data available – many steps short of a full risk assessment. For the 110 'priority' chemicals identified eight years ago under the EU 'existing substances' regulation, no actual action has yet been taken as seemingly endless processes and assessment take place. (There are about 100,000 chemicals on the market in the EU so I am sure our descendants will manage the full job before humans become extinct. Maybe.) Greenpeace's campaign in 1999 to have phthalates removed from PVC teethers and children's toys drew heavily on the limits to our knowledge about exposure and its avoidability. There were good indications of potential hazard from phthalates, including reproductive effects and endocrine disruption, liver and kidney effects, asthma and effects of lung function. Damage to human health could not be proved in any of these cases; such studies would take a considerable amount of time and expense. Thus Greenpeace converted the question 'What is a safe dose of phthalates?' to the question 'Should babies and toddlers be exposed to hazardous chemicals?' The response from consumers and many politicians was a resounding 'no' to the second question and the latter have put an emergency ban on phthalates in teethers. A demonstration of the inability of members of the scientific community to cope with an acknowledgement of ignorance is that the relevant European scientific committee is still trying to find methods of determining how much phthalates will leach out of teethers under the action of saliva and recently reported progress on establishing methodologies for one test. The chemical industry is reportedly 'jubilant' that this is continuing.

Additionally, the endpoints for scientific scrutiny of environmental impacts are actually subject to interpretation, hard to quantify and hard to relate in a cause and effect manner sufficiently unequivocal for the threshold for regulatory or legal action to be reached. If, for example, the endpoint for impact of a particular chemical is to look at the population of a species, the necessary data to establish such an impact would need to include at least the following: emissions data, environmental concentration, epidemiology on the population of the target species compared to a 'similar' population without the chemical stress, plausible biochemical mechanism, trend data over a series of years. It is a moot point as to who is going to have enough interest and cash to resource this kind of research programme. As was pointed out by a speaker at the UNESCO world conference on science on 1999: 'Companies will not do research that benefits campaigners.'

All this demonstrates the limitations of conventional scientific assessment of environmental and health hazards. These impacts have emerged in spite of assurances that operations of chemical and nuclear plants were 'safe' or posed no risk. Judgements implicit and inherent in the regulatory process, those of assimilative capacity, dilute and disperse chemical by chemical assessment are all found to be deeply lacking. Yet this approach is still described (at least by the chemical and nuclear industries) as 'the scientific approach'. But where in science does it say this is the right way to do things? The Department for Looking at Things One at a Time? The Faculty for Missing the Big Picture? Failing to allow for ignorance and uncertainty is, in fact, deeply unscientific and the failure to acknowledge and deal with these thorny issues in favour of pursuing a 'scientific' approach is highly corrosive on public trust in science, particularly when associated with industrial interests.

The wonder is why significant parts of the scientific community and its leaders allowed itself uncritically to accept and continue to support judgements and assumptions so convenient to industrial interests. To find that public trust in science-based decision making is declining is scarcely surprising. It is a wonder it has not declined more. The failure by scientific and policy institutions to acknowledge and take responsibility for ignorance drives the mismatch between their and the public's views; and that public view has a solid empirical basis. Those who continue to advocate a narrow 'scientific' approach do the credibility of science more harm than any number of astrologers and crystal theorists.

ESTABLISHMENT REACTIONS TO PUBLIC MISTRUST

Establishment reaction to the furores created by this mismatch has been to propose remedies for regaining trust in science. These fall, in my experience, into four categories:

- We need to educate the public about the nature of science and the uncertainties.
- Its all the result of an hysterical media.
- It's all the fault of the pressure groups.

- We need more transparency and openness.

Let's examine each in turn.

WE NEED TO EDUCATE THE PUBLIC ABOUT THE NATURE OF SCIENCE AND THE UNCERTAINTIES

This is a modification of the now-discredited public understanding of science movement. Public understanding of science was hopelessly over-optimistic in its aims of raising the general knowledge about science in the public at large. And in any case the idea that a little physics education would make everyone love nuclear power seems bizarre. However, this idea runs, if people were taught what science was really like, uncertain and provisional at the cutting edge, instead of the monolithic certainties that characterize school teaching of science, then we would be saved the 'overreaction' that we saw in the case of GM food. However, as the Economic and Social Research Council Global Environmental Change Programme – a ten-year programme looking at, among other things, public attitudes – concluded when looking at the GM food debate:

> The evidence from research is that many of the public ... [are] highly sophisticated in their thinking on the issues. Many 'ordinary' people demonstrate a thorough grasp of issues such as uncertainty: if anything, the public are ahead of many scientists and policy advisors in their instinctive feeling for a need to act in a precautionary way ... to assume that the public is ignorant and gullible is not only patronising, but inaccurate and therefore damaging to the debate ... People may not know the scientific and technical detail, but they have developed a sharp awareness of the broad issues involved ... in particular, the public mistrusts the scientific approach to ignorance – unknown factors that may lead to 'surprises' in future.

And let us dismiss the notion that the public is anti-science. I see no evidence for this and a great deal to the contrary. Television programmes about dinosaurs or astronomical revelations are still able to command strong viewer figures. Equally, events that offer to debate science and society issues often command hugely impressive turnouts. Instead there is considerable public scepticism over claims coming from science *with a purpose* – where the science being done is geared to a particular policy outcome or furtherance of sectional interests/product performance and safety.

IT IS ALL THE RESULT OF AN HYSTERICAL MEDIA (WHO NEED TO BE CONTROLLED OR 'GUIDED')

Greenpeace routinely tracks expressed public opinion in order to monitor expressed levels of environmental concern. Our tracking data, conducted by MORI, showed 'concern' about GMOs rising from around 20 per cent early in the early 1990s to 70 per cent in December 1998. In February and early March there followed an intense period of media coverage of the GM issue focusing on food safety, environmental and political/commercial issues. At the end of that period Greenpeace commissioned a further tracking study to see what effect such intense media scrutiny had on expressed opinions. The April 1999 level of concern was 71 per cent, within the margin of error (three per cent) for the sample size. Thus, this period of intense coverage had no real effect on whether people expressed concern or not. People did not read the sometimes raucous tabloid headlines and gullibly form an opinion in sympathy with them. Previous opinion research had shown considerable *latent* concern about GMOs in 1996. The result of the media coverage was not to change that level of concern, but to legitimize its expression, which largely took the form, for individuals, of pressuring food retailers and manufacturers to remove GM material from the food supply chain.

However, the important lesson is that underlying opinions were not really affected by the intense media coverage, however partisan, on both sides. This makes the various codes of conduct to guide relationships between scientists and journalists being suggested by scientific bodies all the less justifiable. Indeed, if they act as a way of papering over fundamental differences between official policy and public sentiment, one can consider them actively counterproductive for the governance of scientific issues by society at large. They are also deeply illiberal. As Ian Hargreaves and Galit Fergusson have pointed out:

> The code's instruction to journalists that they should discover and reflect 'the majority view' on issues of scientific controversy, rather than balancing reports by setting a majority view alongside one 'held by only a quixotic minority of individuals' [is unhelpful]. This not only sounds unrealistic in the fast-moving news game, it is also potentially positively dangerous. The point of public debate in democracies is that minority views achieve exposure, not only because they sometimes turn out to be right ... [the Code says that journalists] 'should be wary of regarding uncertainty about a scientific issue as an indication that all views, no matter how unorthodox, have the same legitimacy.' This may be a well-intentioned appeal for journalists to show better judgement. But imagine the same point made about, say, politics or economics: journalists should avoid granting legitimacy to the views of the Green Party or Orthodox Jewry.
>
> (Who's misunderstanding whom?, Economic and Social Research Council, September 2000)

SCIENCE: Knowing about ignorance

IT IS ALL THE FAULT OF THE PRESSURE GROUPS

Naturally one would not expect a member of Greenpeace to say pressure groups have no influence. Contrariwise, those who suggest that pressure groups whip the public into hysteria about various things such as GMOs had better remember one thing: BSE. No pressure groups had any real influence on the outcomes. Nonetheless, it remains a very significant influence on the breakdown of trust between public and government and science. As documented for the Parliamentary Office of Science and Technology, what characterized the GM farrago in particular was the drift in sentiment between public values about food production and official policy on the other. This gives pressure groups like Greenpeace, whose base of influence is public concern, raw material to work with, as agents in civil society who hold powerful institutions to account. But these sentiments can rarely, if ever, be generated from nowhere by campaigning activity. Greenpeace in the UK has actually poured far more resources into campaigning on climate change compared to GMOs in the last five years, with considerably less joy.

WE NEED MORE TRANSPARENCY AND OPENNESS

This idea is different from the three previous ones in that following this line of thought leads potentially to real improvements in decision making, while the others are at best blind alleys or even counter-productive. However, one needs to be clear about the limitations of this approach. More openness in decision making, as a remedy to lack of trust, assumes that when people are allowed to look into the decision process, they will like what they find. As I have argued throughout, the real problem of lack of trust arises from different understandings and appreciations of the uncertainties and ignorance around new innovations, how that ignorance should be handled and the benefits that those innovations provide. These differences of view will not be mitigated by having greater transparency. Sports fans may have noted that increased openness and transparency given to the decisions by football referees and cricket umpires by television cameras and action replays have not increased the trust in those decisions. Where such an approach could have real benefits is that the assumptions about the uncertainties and ignorance can be pulled out and tested against different perspectives. Empirical evidence and political judgements can be teased apart and challenged. So this can have real benefits, but not that this will make it an easier road for the science policy process. Prepare for a rocky ride.

 **PEER REVIEW**

Can the scientific community police itself on unstated assumptions? Probably not. The scientific defence against 'bias' or not being able to acknowledge uncertainties is meant to be the old tradition of peer review where other scientists assess scientific work before its publication. Leaving aside the fact that most data submitted for approval to government committees, for example, on GMOs or pesticides, have not been peer reviewed, this protection may not be what it seems. For one thing, some studies have suggested that expert judgement isn't as good as it might first appear. For another, within each academic scientific discipline similar attitudes to the unknowns in the discipline are likely to prevail because of common training, research methodologies and research foci. This comes about because of the necessary focus for making progress on research and through the formation of disciplinary 'walls', ranging from which journals will publish particular research through to career progression within academic institutions. 'People are "soaking up" views from other members of the group, which can strengthen their own views; and this is a mutually reinforcing process,' explains psychologist Nick Chater from the Institute of Cognitive Science at Warwick University (*The Guardian*, 29 January 2001). This all means that, as Emeritus Professor of Physics John Ziman puts it: 'Basic scientific knowledge is typically fragmented into little islands of near conformity surrounded by interdisciplinary oceans of ignorance' (*Nature*, 382, 29 August 1996). This near conformity means that the peer review process may not be much use, particularly when applying to ignorance about consequences of new scientific discoveries, especially outside disciplinary boundaries. This was no doubt informing the view put forward by the editor of *The Lancet* in saying that peer review was a:

quasi-sacred process that helps to make science our most objective truth teller... But we know that the system of peer review is biased, unjust, unaccountable, incomplete, easily fixed, often insulting, usually ignorant, occasionally foolish, and frequently wrong .

(*Medical Journal of Australia*, 21 February 2000)

CONCLUSION

So what is the way forward in a situation where, as Professor Michael Gibbons describes it, we are in conditions of:

The expiry of the social contract between science and society that has dominated [the period since the Second World War]. A new social contract is now required. This cannot be achieved merely by patching up the existing framework. A fresh approach – virtually a complete 'rethinking' of science's relationship with the rest of society – is needed.

(*Nature* supplement, 2 December 1999).

What would this new arrangement look like? There is no simple solution but there is a minimum of three strands:

1 As science becomes closer to industrial interests so it needs to open up to society as well. At present the publicity organ of the Biotechnology and Biological Research Council is called BBSRC Business. The statement of intent is clear. But why not BBSRC Society or BBSRC for People? There needs to be meaningful interaction between technical experts and those to whom the science-in-policy is being addressed. These communities include farmers, health workers, civil society groups, people local to site-specific issues and others. Anyone, in fact, who feels they should be involved or have an interest. Such people can have different perspectives that shed genuine light on the problems that science seeks to address in the world outside the laboratory and to produce what some academics have called 'socially robust knowledge'.

2 The acknowledgement of ignorance means not just accepting that there are risks which are not yet understood but that these risks are unknowable. Science on its own cannot deal with the risks, even if the wider participatory process of problem solving might help to identify additional issues. The inevitable presence of unknowable risks means that new innovations need a wider mandate than that some company product manager thinks it is a moneymaker. If one accepts – and I for one, certainly do – that science and technology innovation will be one of the most powerful forces shaping the future of society, by what right are the research agendas and technology choices being decided in closed rooms of specialists? Some real choices are already apparent: Will our agriculture be ever more intensified or become more locally based and 'organic'? Will our future energy supplies be based on nuclear or renewable power? These are crucial questions for the future and science and technology agendas will have an

enormous impact on our ability to deliver desired outcomes. Such a wider mandate for scientific agendas can start with opening up scientific innovation processes – although limited, initiatives to consult on research agendas by, for example, NERC are welcome. But much more needs to be done.

3 Changes in the base of science funding has come as national governments provide less and less of the proportion of overall research spend, especially in areas that are seen to have huge commercial potential such as biological science. Although many companies have become global, enhanced communication means that international civil society accountability mechanisms are emerging too. Industrial and brand reputation is emerging as a key driver for technology and product development. Governments, particularly those in large markets like the UK, can encourage this trend because governments can determine a corporation's reputation too. The clever companies will get ahead and realize that they have to engage with people not just as consumers but as citizens, and as citizens people want to see companies not just doing things right by customer care standards, but doing the right things. Engaging with people's hopes and fears about the future of scientific developments is not just good corporate behaviour; it is actually enlightened self-interest. Governments may not be able to direct, but can facilitate this trend.

This is not a complete solution, but the adoption of these measures would perhaps, repromote science to the esteem it merits as an unparalleled generator of reliable (if sometimes rather narrow) knowledge. However, the scientific community will need to work hard to regain that respect and only by doing so will it show that it deserves it.

AFTERWORD
Tony Gilland

This book has highlighted three key issues in discussion of scientific expertise:

- Is scientific objectivity possible?
- How should we respond to the existence of uncertainty and risk?
- Which experts should advise on and inform decisions about new or controversial technologies and should more non-scientists be included in the process?

IS SCIENCE OBJECTIVE?

The extent to which science can provide us with an objective understanding of the world is crucial to our understanding of the role of scientific experts. If science is just another viewpoint or way of looking at the world there is no reason for it to occupy the esteemed position that it traditionally has done. The question of scientific objectivity arises both in relation to the pursuit of scientific knowledge and the application of that knowledge through technological innovation.

One perspective on the pursuit of scientific knowledge is to see science as an open-ended exploration of the physical world that over time provides us with a better explanation of how things are, that is a closer and closer approximation to the truth. From this perspective the relative truth of any piece of scientific knowledge can be judged by assessing how well it can explain and to an extent predict observable phenomena (whether these are observable to the naked eye or through the use of complex equipment). New scientific theories that contradict earlier ones, previously regarded as our most reliable explanation of some aspect of physical reality, can be adopted without undermining our belief in the rationality of the scientific approach. The new theory can be read as a closer approximation to the truth than the previous one which itself might be surpassed by an even more reliable explanation of reality as our understanding of the world improves. The existence of scientific uncertainty is entirely consistent with a belief in the objective character of scientific knowledge. Rather than being an individual exercise open to subjective biases, it is argued that science is a collective endeavour that involves people with specialist knowledge scrutinizing one another's work and that this process of peer review is central to science's claim to objectivity.

An alternative perspective on this question is to view scientific knowledge as inherently unreliable due to the subjective inclinations of scientists shaped by the social world within which they operate. This argument was particularly influential in the work of a number of feminist and environmentalist thinkers in the 1970s and 1980s who argued that the notion of scientific objectivity was little more than a social convention that helped to maintain existing power relationships within society – for example, male and/or western domination in the world. Others have also argued that the ability of scientists to hold each other's work to account is limited by conformity within the scientific community. For example, common training and research methodologies or the pressures on scientists to achieve results, are seen as generating a climate of conformity within particular research areas that makes it difficult for anyone to challenge the prevailing consensus and therefore properly test the accuracy of the knowledge that is being generated. Logically, the adoption of an extreme relativistic position on knowledge denies the very possibility of actually knowing anything about the world; all that we can know is how we as particular individuals see and understand things. This argument is difficult to sustain and, therefore, more realistically, the question becomes one of how much weight to attach to the argument that scientists are swayed by their subjective inclinations in their pursuit of the truth and how suspicious we should be of the ability of the broader scientific community to properly hold the work of individual scientists to account.

TECHNOLOGICAL INNOVATION, UNCERTAINTY AND RISK

Similar issues in relation to the question of scientific objectivity arise when we move from the laboratory to introducing new technologies in society (or, indeed, assessing the safety of existing technologies). However, in this very different context, the issues and arguments are distinct.

Assessing the safety and the benefits of a technology employed or to be employed by society brings additional layers of complexity to the discussion. Clearly the questions to be addressed are not only scientific ones. Questions about the safety of cars or their emissions, for example, are amenable to scientific investigation, while questions about how many roads should be built and where require social and political debate to reach agreed solutions. And, if there is a degree of uncertainty over the implications of different levels of car emissions for human health, then the question of how cautious we should be in the face of that uncertainty is also one that is open to non-scientific debate. Furthermore, how a technology will be employed by society over time is not easy to predict at its point of introduction. When the combustion engine was invented no one could have predicted the high level of car usage in advanced countries in 2001, neither could they have assessed the implications of this for human life. To reach a viewpoint on such discussions, two related issues need to be considered: 1) what science can and cannot tell us and 2) how cautious should we be in the face of uncertainty, in its different aspects.

The question of what science can and cannot tell us is informed by both how trustworthy we believe scientific knowledge to be and by how much importance we attach to what is currently known in contrast to what is not known. For example, in relation to the debate about the safety of GM foods for human health, the majority of scientists with specialist knowledge in this field argue that all the available evidence shows that they are as safe to eat as conventional food products. However, the possibility of some unforeseen impact of eating genetically modified foods on human health can never be ruled out completely – it is logically impossible to prove that something unforeseen will not happen in the future. How then, should we proceed?

One position to adopt is to argue that the scientific knowledge available is itself unreliable because of the subjective inclinations of scientists. For example, it could be argued that scientists funded by government and industry who, to different degrees, are regarded as favourable towards the use of GM foods are likely to be biased towards the introduction of GM foods for fear of losing their funding. Therefore they may interpret the available evidence in an overly positive way and overlook potential pitfalls. It could also be argued that scientists who have devoted many years of research to genetic engineering are keen to see that research put to beneficial use by society, for either altruistic and/or egotistical reasons, and that this also introduces a bias into their scientific assessments of the technology.

Finally, even if we play these arguments down and assume that the knowledge produced by scientists is robust and reliable, an argument for adopting a cautious approach towards GM foods can still be made on the grounds that more weight should be attached to what we do not know than what we do. This argument rests on the idea that because the natural world is very complex, however much we might know about a new technology, there will be many possible problems that we just cannot foresee and that, therefore, we should be cautious.

In opposition to these views, there are a number of arguments that can be made in addition to those outlined in the previous section. It can be argued that scientists, through their expert knowledge, are the people best placed to know what the most important questions to ask about the introduction of a new technology are and to investigate. It is not that they ask narrow questions and overlook potential pitfalls, but rather that they know when a theoretical risk is just that and is implausible and therefore an unhelpful line of questioning that will shed no light on the task in hand. Finally, while unforeseen events can and do occur, it is important to embrace such uncertainty because it is in so doing that society creates new and beneficial opportunities that improve our lives and help us to cope with adverse circumstances in the future.

WHICH EXPERTS DO WE NEED?

If one accepts the argument that modern society should adopt a more cautious approach to its interaction with the natural world, it is not strictly necessary to argue for additional and new types of experts to inform official decision-making processes. It would be possible to maintain that scientists are best placed to advise government and other bodies as to what is known about a technology, but at the same time to campaign and lobby for politicians and others to adopt a more precautionary approach in decision making, and to insist that those who wish to introduce a technology need to provide further scientific evidence that no harmful consequences are likely to result.

However, if we combine the argument for a more precautionary approach with the argument that scientific expertise has inherent limitations, this raises the question of whether the inclusion of other viewpoints or types of expertise in the advisory process itself would improve the quality of information available to decision-makers. If, for example, it is accepted that scientists have a tendency to adopt too instrumental and arrogant an approach towards the relationship of human beings to the natural world, then it could be argued that the involvement of environmentalists within the advisory process will help to ensure that the scientific experts at least consider questions that they might otherwise have dismissed or paid little attention to. Similar arguments could be made for the need to involve consumer experts or ethicists to ensure that scientists do not ignore questions of concern to the wider public that, from the viewpoint of the scientific specialist, appear to be of little significance.

Alternatively, if we believe that scientific experts are best placed to know which are the most important questions to ask, and how to go about answering them, then the benefit of including additional non-scientific viewpoints within official advisory systems is far less obvious. From this perspective, we could adopt broadly one of two attitudes towards the new forms of expertise that have been discussed in this book. One could argue that the inclusion of non-scientific experts is irrelevant but not particularly harmful and that if it improves some people's confidence in the regulation of technology then overall it may still be a good thing. Alternatively, one could argue that the inclusion of non-scientific viewpoints at this stage in the process confuses scientific discussions with ostensibly political discussions by raising irrelevant or even obstructive questions.

One thing that is certain is that the question of how to improve society's faith in official decision-making and regulatory processes, when it comes to issues related to science and technology, is a high priority. How society views experts and expertise is clearly central to shaping our view of the best way forward. I hope that this book has helped the reader to develop his or her own views on these questions.

DEBATING MATTERS

Institute of Ideas
Expanding the Boundaries of Public Debate

If you have found this book interesting, and agree that 'debating matters', you can find out more about the Institute of Ideas and our programme of live conferences and debates by visiting our website **www.instituteofideas.com**. Alternatively you can email **info@instituteofideas.com** or call 020 7269 9220 to receive a full programme of events and information about joining the Institute of Ideas.

Other titles available in this series:

DEBATING MATTERS

Institute of Ideas
Expanding the Boundaries of Public Debate

DESIGNER BABIES:

WHERE SHOULD WE DRAW THE LINE?

Science fiction has been preoccupied with technologies to control the characteristics of our children since the publication of Aldous Huxley's *Brave New World*. Current arguments about 'designer babies' almost always demand that lines should be drawn and regulations tightened. But where should regulation stop and patient choice in the use of reproductive technology begin?

The following contributors set out their arguments:

- Juliet Tizzard, advocate for advances in reproductive medicine
- Professor John Harris, ethicist
- Veronica English and Ann Sommerville of the British Medical Association
- Josephine Quintavalle, pro-life spokesperson
- Agnes Fletcher, disability rights campaigner.

TEENAGE SEX:

WHAT SHOULD SCHOOLS TEACH CHILDREN?

Under New Labour, sex education is a big priority. New policies in this area are guaranteed to generate a furious debate. 'Pro-family' groups contend that young people are not given a clear message about right and wrong. Others argue there is still too little sex education. And some worry that all too often sex education stigmatizes sex. So what should schools teach children about sex?

Contrasting approaches to this topical and contentious question are debated by:

- Simon Blake, Director of the Sex Education Forum
- Peter Hitchens, a columnist for the *Mail on Sunday*
- Janine Jolly, health promotion specialist
- David J. Landry, of the US based Alan Guttmacher Institute
- Peter Tatchell, human rights activist
- Stuart Waiton, journalist and researcher.

ART:

WHAT IS IT GOOD FOR?

Art seems to be more popular and fashionable today than ever before. At the same time, art is changing, and much contemporary work does not fit into the categories of the past. Is 'conceptual' work art at all? Should artists learn a traditional craft before their work is considered valuable? Can we learn to love art, or must we take it or leave it?

These questions and more are discussed by:

- David Lee, art critic and editor of *The Jackdaw*
- Ricardo P. Floodsky, editor of artrumour.com
- Andrew McIlroy, an international advisor on cultural policy
- Sacha Craddock, an art teacher and critic
- Pavel Buchler, Professor of Art and Design at Manchester Metropolitan University
- Aidan Campbell, art critic and author.

ALTERNATIVE MEDICINE:

SHOULD WE SWALLOW IT?

Complementary and Alternative Medicine (CAM) is an increasingly acceptable part of the repertory of healthcare professionals and is becoming more and more popular with the public. It seems that CAM has come of age – but should we swallow it?

Contributors to this book make the case for and against CAM:

- Michael Fitzpatrick, General Practitioner and author of *The Tyranny of Health*
- Bríd Hehir, nurse and regular contributor to the nursing press
- Sarah Cant, Senior Lecturer in Applied Social Sciences
- Anthony Campbell, Emeritus Consultant Physician at The Royal London Homeopathic Hospital
- Michael Fox, Chief Executive of the Foundation for Integrated Medicine.

COMPENSATION CRAZY:

DO WE BLAME AND CLAIM TOO MUCH?

Big compensation pay-outs make the headlines. New style 'claims centres' advertise for accident victims promising 'where there's blame, there's a claim'. Many commentators fear Britain is experiencing a US-style compensation craze. But what's wrong with holding employers and businesses to account? Or are we now too ready to reach for our lawyers and to find someone to blame when things go wrong?

These questions and more are discussed by:

- Ian Walker, personal injury litigator
- Tracey Brown, risk analyst
- John Peysner, Professor of civil litigation
- Daniel Lloyd, lawyer.

NATURE'S REVENGE?

HURRICANES, FLOODS AND CLIMATE CHANGE

Politicians and the media rarely miss the opportunity that hurricanes or extensive flooding provide to warn us of the potential dangers of global warming. This is nature's 'wake-up call' we are told and we must adjust our lifestyles.

This book brings together scientific experts and social commentators to debate whether we really are seeing 'nature's revenge':

- Dr Mike Hulme, Executive Director of the Tyndall Centre for Climate Change Research
- Julian Morris, Director of International Policy Network
- Professor Peter Sammonds, who researches natural hazards at University College London
- Charles Secrett, Executive Director of Friends of the Earth.

310579

THE INTERNET:

BRAVE NEW WORLD?

Over the last decade, the internet has become part of everyday life. Along with the benefits however, come fears of unbridled hate speech and pornography. More profoundly, perhaps, there is a worry that virtual relationships will replace the real thing, creating a sterile, soulless society. How much is the internet changing the world?

Contrasting answers come from:

- Peter Watts, lecturer in Applied Social Sciences at Canterbury Christ Church University College
- Chris Evans, lecturer in Multimedia Computing and the founder of Internet Freedom
- Ruth Dixon, Deputy Chief Executive of the Internet Watch Foundation
- Helene Guldberg and Sandy Starr, Managing Editor and Press Officer respectively at the online publication *spiked*.